A GEOGRAPHY OF
soils

THE BROWN
FOUNDATIONS OF GEOGRAPHY
SERIES

Consulting Editor
ROBERT H. FUSON
University of South Florida

A GEOGRAPHY OF

Agriculture
James R. Anderson, University of Florida

Soils
Robert M. Basile, University of Toledo

Transportation and Business Logistics
J. Edwin Becht, The University of Oklahoma

Plants and Animals
David J. de Laubenfels, Syracuse University

Geography
Robert H. Fuson, University of South Florida

The Atmosphere
John J. Hidore, Indiana University

Population and Settlement
Maurice E. McGaugh, Central Michigan University

Industrial Location
E. Willard Miller, The Pennsylvania State University

Water
Ralph E. Olson, The University of Oklahoma

Earth Form
Stuart C. Rothwell, University of South Florida

Landforms
John H. Vann, California State College, Hayward

Minerals
Walter H. Voskuil, University of Nevada

THE BROWN
FOUNDATIONS OF GEOGRAPHY
SERIES

A GEOGRAPHY OF
soils

ROBERT M. BASILE

The University of Toledo

WM. C. BROWN COMPANY PUBLISHERS
DUBUQUE, IOWA

**THE BROWN
FOUNDATIONS OF GEOGRAPHY
SERIES**

Consulting Editor
  ROBERT H. FUSON
  University of South Florida

Copyright © 1971 by
Wm. C. Brown Company Publishers

ISBN 0–697–05151–X

Library of Congress Catalog Card Number: 70-146327

Printed in the United States of America

Preface

This book is intended primarily for the first course in physical geography at the college and university level. It is also suitable for the earth science curriculum where a better understanding of soils of the world is desirable. It will prove useful to other geographers for ready reference.

The many facets of soil science, although touched upon, have not been stressed unduly. However, those basic factual materials thought necessary to a better understanding of the geographical aspects have not been slighted.

The newer classification—the 7th Approximation—was published in 1960 and has been in use in the soil surveys in the United States since then. Since the terminology of the older system is well entrenched in the literature, it was thought advisable to cross reference the two systems as both will be in use for some time to come. Both systems are discussed. More complete discussions may be obtained from the bibliographic references.

It is suggested that local soil survey reports be obtained by the student for closer and more detailed study of the more familiar home area.

Geography is one of man's oldest sciences, yet it is as new as the Space Age. Knowledge of the earth obtained from satellite photography and measurement, remote sensing of the environment, and by means of other sophisticated techniques are really but a stage in the evolutionary process that began with ancient man's curiosity about his surroundings. Man has always been interested in the earth and the things on it. Today this interest may be channeled through the discipline of geography, which offers one means of organizing a vast amount of physical and cultural information.

The **Brown Foundations of Geography Series** has been created to facilitate the study of physical, cultural, and methodological geography at the college level. The **Series** is a carefully selected group of titles that covers the wide spectrum of basic geography. While the individual titles are self-contained, collectively they comprise a modern synthesis of major geographical principles. The underlying theme of each book is to foster an awareness of geography as an imaginative, evolving science.

Contents

Introduction

The thin veneer of loose unconsolidated materials covering the surface of the earth that we call soil is made up of materials that have had their origin in a variety of flora, fauna, and parent rock materials. Mechanical and chemical weathering processes break down the lithosphere into fragments which, although not a true soil, are the stuff of which soils are made; and the flora and fauna within this material modifiy it further.

Although the same kinds of rock materials are to be found in widely scattered areas of the world from the tropics to the poles and from mountain to plain, the soil formed thereon is not necessarily the same, even though the parent materials may be identical in origin. Differences also exist between those "soils" on disturbed and relocated materials on Indian mounds, those found in close proximity to those mounds, and the soils of tilled fields of forested regions from those soils of grassland regions. All environments are not identical and dissimilar soils result.

Soil Science has evolved over the years as investigators studied the earth surface in widely separated areas under widely varying conditions. Soil scientists have organized their findings into a body of facts dealing with the origin and description of characteristics of soils. They have also attempted to arrange them into groups in which the first few inches or first few feet below the surface have been described. Soil Science, or Pedology, treats soil as a natural body and, with little emphasis on uses of the soil beyond its agricultural implications, sets about determining the characteristics and genesis of the soil by systematic scientific investigation. The chemist, physicist, biologist, zoologist, and the applied scientists (such as the engineer and physical geographer) have all contributed to this body of knowledge. As

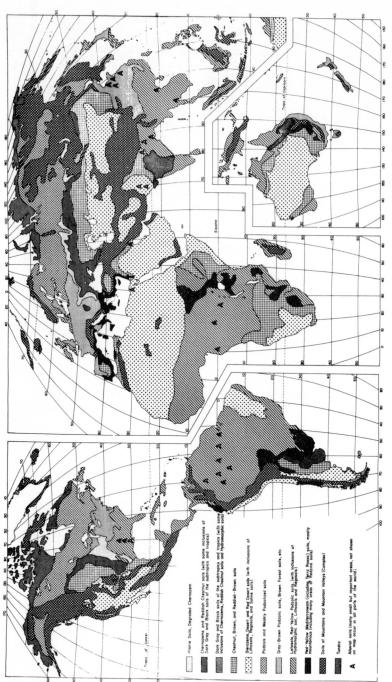

Figure 1.1. Soil map of the world. A relatively recent compilation of the soils of the world using the system of classification in use in the United States as well as world wide prior to 1960. The terminology of this classification is at the present much more familiar to the geographer than that of the 7th Approximation now being used by the soil scientist in the classification of soils in the United States. Courtesy U.S. Soil Conservation Service.

2

with all sciences, classification eventually became a part of Pedology, too. With classification the mapping of the spatial distribution of like soils is an easy step. Figure 1.1 shows a relatively recent compilation of the primary groups of soils of the world. Because of its small scale the map is considerably simplified from reality.[1]

A comparison of world soils with Figure 1.2 (a map of the distribution of the natural vegetation of the world) shows close resemblance on this scale. Note the closely similar boundaries between the Prairie (Brunizem) Chernozem and Chestnut soils, and the prairie lands in temperate regions. Also note the similarity of the boundaries of the Podzols on the soils map and those of the coniferous forests of the United States, Canada, and Alaska. A similar relationship may be seen by comparison of soil maps and climate maps (Figure 1.3). A

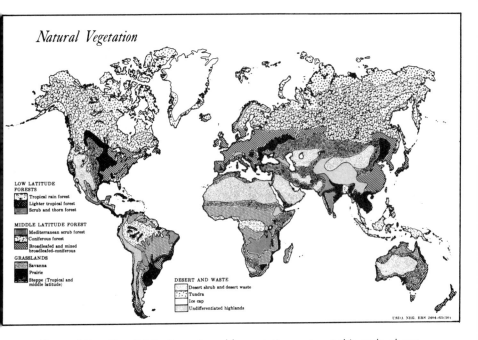

Figure 1.2. The distribution of world vegetation even at this scale shows a close correlation of specific great soil groups. Note the close similarity of boundaries between the Prairie and Chernozem soil with the boundaries of the Prairie grasslands; or the similarity of boundaries between Sierozem soils and Desert shrub. Courtesy U.S.D.A.

---

[1]A still more recent map of world soils is that shown in Figure 5.2. Much of this, however, is conjecture since the major work in soils classification under the new system is being done in the United States.

study of the distribution of soil types has inherent within it, and may be considered as, a synthesis of climate and vegetation.

Man uses soils for many purposes, consequently he studies them with a specific purpose in mind. The highway engineer, interested in the load-bearing capabilities of a soil, has an entirely different set of values in studying a particular soil area than does the farmer who tills that soil for specific crops. The irrigation engineer is concerned with losses of water by seepage downward through the bottoms of his canals and might be very happy with a soil having an impervious layer which would greatly reduce seepage in the subsoil. The farmer would probably abandon such a soil, if under cultivation, or he might never plow it in the first place because of its obvious restrictions to plant root development. The farmer might find an organic soil such as muck ideal for vegetable crops, while the civil engineer would shun the area entirely as an extremely poor place over which to construct a highway or upon which to build some structure.

A soil geographer, as opposed to the soil scientist, is more concerned with the distribution and morphology of soils types; hence he is more interested in the classification of soils and the observable soil characteristics themselves than the purely scientific researches in the areas of the soil chemistry, soil physics, and the like. Soil morphology, more than anything else, synthesizes the findings of many closely related disciplines in soil profile descriptions.[2] The soil profile is basic in any classification of soils. This is not to say the geographer dismisses the work of the soil chemist and soil physicist from his studies. Once the classification and distribution (the "what" and "where") of soils are accomplished, the soil geographer inquires as to "why" they are and even goes beyond that to investigate "how" man may make the best use of the soils of his environment.

Consequently, soil geography can be considered as a study of the distribution of the soils of the earth, their characteristics, and how they developed. In addition, soil geography is concerned with how soil might best be used and not abused. Of foremost importance to the rapidly increasing world population are those soils capable of food production not yet used or not fully utilized for that purpose. Table 1.1 summarizes by soil groups the potentially arable land available throughout the world. At present less than half of the world's potentially arable land is cultivated. South America and Africa contain the largest amounts of such land. Europe and Asia have a large percentage of their arable land already under cultivation.

---

[2]See Appendix for definition of soil profile.

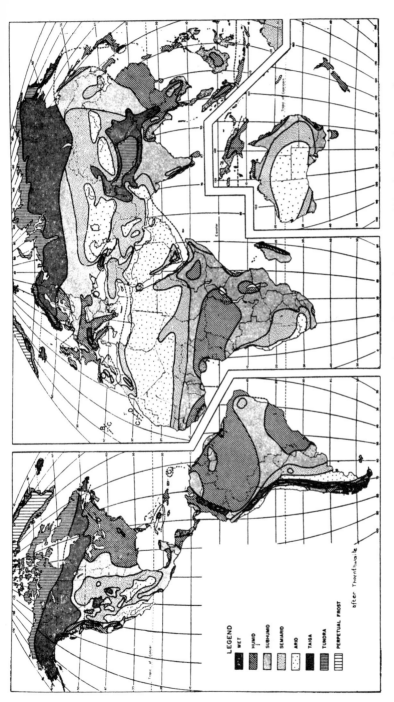

Figure 1.3. Climate map of the world. The climates of the earth according to C. W. Thornthwaite correlate more closely with some vegetation types and soil groups but not all. While temperatures are only vaguely implied there is evidence of temperature change with the changing character of the forest (and soils) in the humid climatic areas. Other similar comparisons may be made. Courtesy U.S.D.A.

LEGEND
WET
HUMID
SUBHUMID
SEMIARID
ARID
TAIGA
TUNDRA
PERPETUAL FROST

after Thornthwaite

5

TABLE 1.1

Less than half of the potentially arable land is cultivated today. Much of the vacant land is in South America and in Africa while in Europe and Asia a very high percentage of arable land is now cultivated. (Data from U.S.D.A.)

ESTIMATED POTENTIALLY ARABLE LAND IN THE WORLD

	Potentially Arable Land in Map Unit	
	Percent	Acres (millions)
1. Prairie Soils, Degraded Chernozems	80.0	242
2. Chernozems and Reddish Chestnut	70.0	660
3. Dark Gray and Black Soils of Subtropics and Tropics	50.0	618
4. Chestnut, Brown, Reddish Brown	30.0	892
5. Sierozems, Desert	.5	34
6. Podzols and Weakly Podzolized...........	10.0	320
7. Gray-Brown Podzolic	65.0	972
8. Latosols, Red-Yellow Podzolics	35.0	2,780
9. Red-Yellow Mediterranean	15.0	41
10. Soils of Mountains	.5	30
11. Tundra	.0	0
Total		6,589

These criteria were used to define arable land:

That reasonably good management would be used including appropriate combinations of adapted crop varieties, water control methods, pest control, and methods of plant nutrient maintenance, including some chemical fertilizers.

Crops include the ordinary food, fiber, and industrial crops that are normally cultivated as well as fruits, nut crops, rubber, sisal, coffee, tea, cocoa, palms, vines, and meadow crops that may or may not be cultivated.

All regular fallow land is counted, including the natural fallow under shifting cultivation.

Irrigation of arid soils is limited by water from streams and wells. Sea water is excluded as a potential source.

Some knowledge of the fundamentals of Soil Science is necessary in order to comprehend soil morphology, soil classification, and why soils are mapped in the manner that they are. Toward this end the following discussion deals first with the materials of which soils are made, the time required for them to evolve, the topographic, vegetative, and biotic factors in their development, and the role played by climate in their formation.

Occasionally an undefined term may appear prior to its logical place of definition. The glossary in the appendix will clarify the meaning.

SELECTED REFERENCES

University of East Anglia. *Geographical Abstracts.* University Village, Norwich, England 1966-. Major articles on soils abstracted from the various journals are found in one of the four series published. It is a valuable source of information on soils of interest to the geographer.

Soil Formation, Development, and Properties

The Regolith

While it is not our purpose to discuss weathering of the earth's crust to any great extent, a mention of the process is necessary to understand the formation of the parent material of soils.

Underlying all soils are bedrock of sedimentary, igneous, or metamorphic origin; or recently deposited glacial, wind, or water-laid materials. If a sedimentary rock such as a sandstone (largely quartz grains) were weathered mechanically, it would be broken down into fine fragments of the original materials by such activities as freezing and thawing, abrasion, heating and cooling, wetting and drying. If the bedrock were igneous or metamorphic, the mineral composition would be much more varied and such minerals as the feldspars, hornblende and apatite, (among others) would compose part of the fragments (Figure 2.1). Through such mechanical weathering or *disintegration* the bedrock is fragmented. The small pieces of rock and mineral matter are similar to the original in all respects, but are now separated from one another by *pore spaces* through which air and water may move. Weathered chemically the rock and mineral matter is *decomposed*. The material that remains is unlike the original in that certain chemical elements are altered or removed entirely. Thus calcium, potassium, sodium, aluminum, iron, and many other elements are chemically removed from the mineral and usually exist in combination as oxides, sulfates, carbonates, or silicates as secondary minerals. Weathering is less active beneath the surface and as a consequence the fragmented bedrock is more coarse farther down from the surface. This weathered zone of primary and secondary mineral materials or *regolith* may vary considerably in depth from a few inches to many feet in thickness,

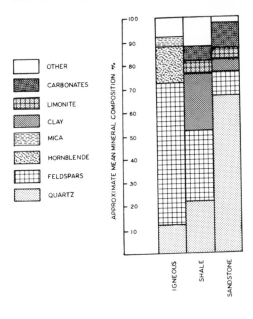

OTHER

CARBONATES

LIMONITE

CLAY

MICA

HORNBLENDE

FELDSPARS

QUARTZ

Figure 2.1. The approximate mean mineral composition of igneous, shale, and sandstone rocks is given on a percentage basis. Sandstones, while variables as to their content of other minerals, nevertheless contain close to 70% quartz. Igneous rocks contain about 80% of the ferromagnesian minerals and about 10% quartz.

depending on the extent of the weathering activity. Figure 2.2 represents such a zone of fragmented, weathered rock 8 or 10 feet in depth above limestone.

At the very top of the regolith, in the area immediately under the plant life and in the general area of the root zone of those plants, is an area of considerably different character than the regolith from which it has evolved. This is the soil or *solum*. The soil while providing a mechanical support for plants ideally should be composed of about 50 per cent of mineral and organic matter and 50 per cent pore space containing both air and water (Figure 2.3).

Rain falling on the surface of the regolith will run off in part, and some will run in and through the solum and regolith to aid in soil development. The percolating water is active in dissolving those mineral portions of the regolith that are susceptible to solution, as well as carrying the load of dissolved mineral matter downward into the regolith. As this *solution* activity continues, rapid changes in the chemical characteristics of the upper portion of the regolith occur. Carbonate and sodium salts, as well as other soluble materials, will be removed from the upper regolith to be later deposited a few inches to a few feet below, if the percolating water has not reached the water table. An enrichment results with the depositions and accumulation of the dissolved load in the zone of accumulation of the upper regolith a few inches to a foot or so beneath the surface. On the other hand, some groundwater reacts to gravitational pull and drains away through the soil, carrying much of the dissolved load with it.

Not only does solution activity occur within the upper zone in the regolith, but the removal of some very fine-sized solids (*colloidal* ma-

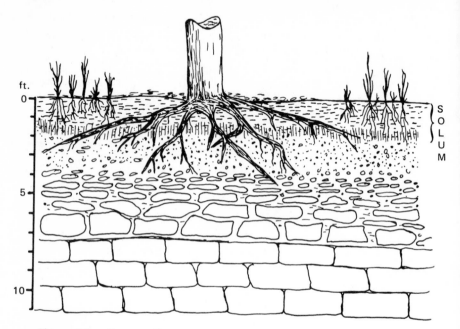

Figure 2.2. The regolith is an area of variable thickness composed of weathered rock materials. The surface few inches to few feet, which is modified by other activities in addition to weathering, is the solum.

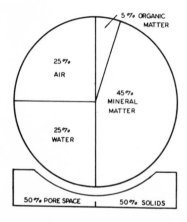

Figure 2.3. The composition of the ideal mineral soil should be about half pore space and half solids. The pore space will be saturated with water after a rain; however on the average it should contain about equal amounts of air and water. The small amount of organic matter aids in maintaining good soil tilth.

terials) such as clay may also result from the downward movement of water within the pore space. This process of moving colloidal materials through the profile is *eluviation,* and results in the surface of the regolith having a higher percentage of the sandy material than it formerly had. As the water continues downward, much of the colloidal material in suspension will be dropped, increasing the content of the finer materials in the lower part of the regolith. This enrichment of the lower layer of the regolith with colloidal materials is the process of *illuviation,* and is a further modification of the original regolith to the extent that a noticeable banding may be observed.

By leaching of soluble materials by eluviation and by illuviation, the original regolith will become layered in its upper few inches or upper dozen feet or so. As with any profile view of a subject, whatever it may be, where it is viewed in its entirety from top to bottom, its character, shape, size, and texture irregularities are revealed. The layers or *soil horizons* are basic to soil identification and collectively are known as the *soil profile.*

Figure 2.4 is a line drawing of how a complete soil profile might look to the observer, although all soil profiles will not have all horizons in evidence. The upper O horizon, for example, may be completely destroyed under cultivation and some other horizons may not be in evidence due to irregularities in, or nonexistence of, leaching, eluviation, and illuviation, among other factors.

Factors in Soil Formation

parent materials

Granitic parent bedrock is rich in minerals containing iron, potassium, and magnesium in addition to the quartz of which many sandstones are composed. Soil water moving through weathered parent material of this type may contain large amounts of dissolved mineral matter and clay, so that a marked horizon development will result from eluviation and illuviation. A regolith developing upon the metamorphic rock quartzite (largely quartz) will vary greatly in mineral content from a regolith developing on gneiss (quartz, feldspars, and ferromagnesium minerals), and the soils developing upon each will vary greatly in mineral content and in horizon development. As one might suspect, parent rock affects the soil that develops upon it. Those soils developed in place upon the bedrock of the region are broadly referred to as *residual soils,* whereas those developing upon materials transported by glaciers, wind or running water are broadly classified as *transported soils.* However, because of differences in weathering, similar soil parent materials may be produced from different geologic materials and unlike parent materials may be produced from similar rocks.

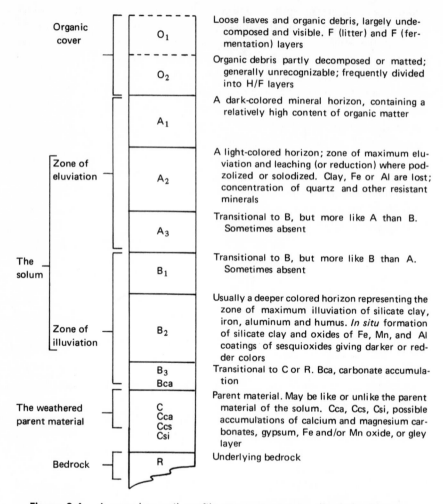

Figure 2.4. A complete soil profile may not contain all of the horizons as listed in this hypothetical case, however, it will contain many of them. The amount of organic material will vary with the kind of vegetation; the zones of eluviation and illuviation will vary in depth and character depending on the amount of water available; accumulations of calcium carbonate, calcium sulfate and silica in the lower B horizon depend to a considerable extent upon the climate.

The major physiographic regions of the United States are outlined in Figure 2.5, together with major complexes of parent materials for soils. From it one can infer major areas of residual and transported regolith as well as the several mineral compositions of those materials. The Piedmont plateau, for example, contains deep residual soils developed largely from metamorphic rocks. The Atlantic-Gulf coastal plain, made up largely of unconsolidated or poorly consolidated sedimentary materials, contains many sandy and silty residual soils. On the other hand, those soils of the Central Lowlands have developed upon till and outwash plains of glacial origin as well as *loess*. (The loess of wind origin for the most part is a very uniform, highly siliceous yellow silt.) These, as well as the alluvium of the flood plains of the major rivers, are parent materials for transported soils.

Soil profile development may be interrupted at any time. A stream load will frequently be deposited upon the floodplain or at the stream mouth as a deltaic deposit. This is a seasonal occurrence. The amount of alluvium deposited may be great in humid regions, and range downward to almost nothing in arid areas. Will these have well-developed soil profiles? Very likely not. Year after year such areas are covered with water during floodstage. Layer after layer of silt and sand is laid down on top of the floodplain or delta destroying or covering any incipient profile that may have begun to develop. Furthermore, newly deposited alluvium may be quite unlike the previous deposition of an earlier flood, resulting in variations in the alluvium from silts and clays to coarse sands and gravels. Although quite heterogeneous, alluvial soils are productive and highly desirable in many areas of the world. Alluvium, colluvium, loess, glacial debris and even marine sediments cover sizeable portions of the world and are in reality the parent materials upon which many soils develop.

Figure 2.6 is a more detailed illustration of the variability of parent materials for soils as they are found in Ohio. The western half of the state lies within the Central Lowlands physiographic province and with the exception of the northern quarter much of its surface is composed of calcareous limestone tills of Wisconsin age. The northern quarter of glacio-lacustrine origin is made up largely of heavy, poorly drained calcareous clays and glacial till. The northern area of the state bordering on Lake Erie consists of glacio-lacustrine materials also; however, in this case it is largely of sandstone and shale origin, some of which is quite acid.

The northern half of the Appalachian Plateau in Ohio has a regolith that is largely of glacial drift derived from the underlying sand-

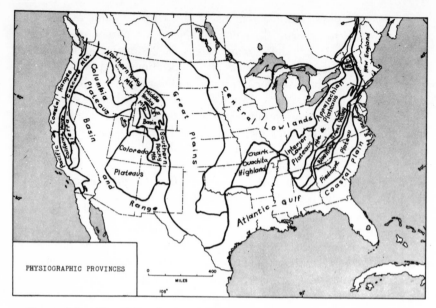

Figure 2.5. The Physiographic Provinces are composed of various kinds of rock and rock materials from which soils develop. Only the major kinds of surface materials found are listed for each of the Physiographic subdivisions of the United States.

New England—mostly glaciated metamorphic rocks.
Adirondack Mountains—glaciated metamorphic rocks.
Appalachian Mountains and Plateaus—shales and sandstones and other sedimentary rocks.
Great Valley—limestone for the most part.
Blue Ridge Mountains—metamorphic rocks.
Piedmont Plateau—metamorphic rocks.
Atlantic-Gulf Coastal Plain—horizontal sedimentary rocks, some unconsolidated; alluvium in Mississippi valley and delta.
Ozark-Ouachita-Interior Lowlands—sandstones, shale and limestone.
Central Lowlands—mostly glaciated sedimentary rocks and loess.
Superior Uplands—glaciated metamorphic and sedimentary rocks.
Great Plains—sedimentary rocks.
Rocky Mountains—Steeply folded sedimentary, metamorphic and igneous rocks.
Columbia Plateau—mostly igneous rock; loess in Columbia and Snake valleys.
Basin and Range—gravels, sands alluvium, playa deposits, sedimentary and igneous ridges.
Sierra-Cascade Mountains—Granitic and volcanic crystalline rocks.
Pacific Coastal Ranges—sedimentary rocks.
California Valley—unconsolidated alluvium.

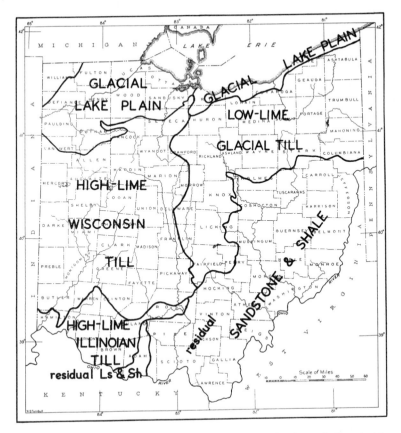

Figure 2.6. Surface materials in the Central Lowlands and Appalachian Plateau of Ohio vary from sandy residual materials to heavy glacial lake plain clays. The soils that have developed within the state show wide differences in profile characteristics.

stones and shales. Both Wisconsin and Illinoian till are represented, consequently a time factor is evident since the tills are of different age.

The southern half of the Ohio portion of the Appalachian Plateau is covered with a residual regolith of sandstone and shale. Being non-glaciated, it is more hilly with many steep slopes and narrow valleys.

One last area, a small tract in the south-central part of the state on the Ohio River contains a nonglaciated surface. However, here the regolith has developed from residual limestone. Like the unglaciated sandstone/shale area adjacent to the east, it is hilly with steep slopes and is badly eroded in many areas.

The recently deposited alluvial materials are less weathered in place than the metamorphic crystalline rocks of the Piedmont Plateau. The fragmentary material making up all alluvium is relatively fresh material devoid of all the products of solution, eluviation, illuviation, and biotic activities found in the well developed soil from which it may have come. Implied here is a *time factor* that must be considered in the development stages of a soil profile. Also implicit is the fact that some local parent materials for soils were transported from some other place, carried by water, wind, or glacial ice, very recently or thousands of years ago. Broadly considered as *transported* soils, these materials might better be referred to as transported regolith since in many cases insufficient time has passed for good soil profile development.

the time factor in soil development

Like alluvial material, fresh glacial till may not show a profile. However, fresh till in the Arctic has been more rapidly modified than thought possible. Research has shown that the depth of the soil increases from a fraction of an inch to a few inches during the first few decades of its existence, to approximately a half foot in the next hundred years.

Jenny, in his book *Factors of Soil Formation*, reports of work showing that within approximately fifty years after the violent eruption of Krakatoa in 1883, the surface volcanic dust was found to be modified to a depth of about 14 inches, below which the parent material of pumice extended to well over 100 feet.[1] On Dutch polders the loss of calcium carbonate from the surface by leaching took approximately 300 years.[2] The change is gradual. Imperceptibly, parent material is altered. Its general homogeneity is modified through time to the extent that eluviation and illuviation have been able to progress. Somewhere in the interval a soil is born. Its profile develops and becomes more pronounced with passing time. The *solum*, or true soil, is recognizable from the AB horizons, the C horizon being parent material.

These developing profiles vary widely from one latitude to another and from coast to continental interior, yet a pattern of distribution emerges which is dependent on the above and other factors. The profile is a signature, its nature molded and modeled by the variety of soil-forming factors. A profile in its broadest aspects will identify a

---

[1]Jenny, Hans, *Factors of Soil Formation*, McGraw-Hill Book Company, New York, 1941, p. 36.
[2]*Ibid.*, p. 44.

broad group of soils, such as the Podzols of intermediate latitudes or the Latosols of low latitudes.

Groups of soils are not continuous. The interruptions sometimes are quite abrupt but most often a broad soil group will grade gradually into another group, the representative profile of which is significantly different. The difference will be noted, not only with topographic position and vegetative cover of the land, but with climatic change as well.

the topographic factor

Topography and slope as factors in soil profile development are shown in Figure 2.7, which illustrates the various soil series found in a relatively restricted area. The Miami series, which developed on a parent material of glacial till, is shown on the diagram of western

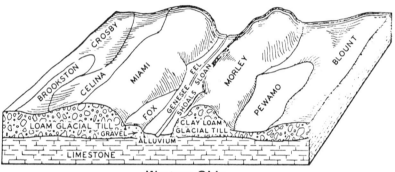

Western Ohio

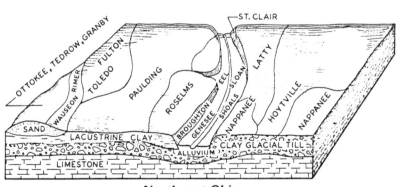

Northwest Ohio

Figure 2.7. Topographic differences even on relatively level land are often sufficiently great to cause minor profile differences. Many soil series may then be identified within a short distance.

Ohio as a soil of the uplands. The infiltration rate of water into a soil and runoff on its surface affects profile development sufficiently so that in a topographically different position the soil develops differently. Even though the parent materials are the same, the Celina and Crosby profiles are enough different from the Miami that the new soil series are delineated. The same may be said for the Brookston series on the bottom slopes of the undulating glacial topography.

Variations in the degree of slope of a hillside affect the rate of water runoff and thus erosion, run-in, and soil moisture retention. In turn these will affect the character of the profile.

Hilltops devoid of vegetation are susceptible to erosion and desiccation by the wind. Very little water will percolate downward through the developing soil. The gradient of the slope below the hilltops affects the rate of runoff as well as the amount. On steep slopes almost all precipitation will run off, whereas much more moisture will percolate into the soil on less steep slopes. Thus the soil forms as a continuum from upland to lowland with gradual changes from one to the other with only occasional sharp differences within narrow areas. This might occur when an undisturbed sandy beach ridge lies adjacent to glacial lake-plain material. In the former the soil profile will be sandy and light colored while in the latter it is likely to be composed of heavy clay and be dark in color. Over great areas marked differences do occur, as they do from humid eastern United States to semiarid and arid western United States and into the high altitudes of mountains. They are not a product of topography, nor time, nor parent materials, but rather a product of climate and vegetation growing upon them.

the climatic factor

As has been stated, soils do vary as the climate varies, altitudinally as well as latitudinally. In the low latitudes under the hot humid conditions of the tropical rainforests, leaching and chemical alteration of the soil continues at a relatively rapid rate. The perpetually high temperatures and abundant precipitation are conducive to rapid chemical weathering and rapid leaching of the developing or developed soil. Not only does weathering proceed at a rapid rate but these processes are also more extreme than in middle latitude temperate regions. In the latter, where periods of glaciation have interrupted decomposition of the rock, depth of weathering is considerably less than in the tropics where hydrolysis, hydration, and oxidation are intense.

Hydrolysis involves the reaction of a host of soil minerals with groundwater and as such is chemical decomposition. Such a reaction may be illustrated with microcline, potassium rich mineral.

$$KAlSi_3O_8 \ + \ HOH \ \longrightarrow \ HAlSi_3O_8 \ + \ KOH$$

The potassium hydroxide formed in the reaction is soluble and may be removed by percolating water or it may be adsorbed on the colloidal complex to be eventually taken up by plant roots. The acid aluminum silicate, $HAlSi_3O_8$, may through further reaction recrystalize to form secondary clay in the subsoil.

Hydration is the combination of water with a mineral to form an integral part of that mineral. Limonite, the yellow hydrated iron oxide, often forms in the soil from hematite, the red oxide of iron.

$$2Fe_2O_3 \ + \ 3H_2O \ \longrightarrow \ 2Fe_2O_3 \ \cdot \ 3H_2O$$

Solution may occur. Limestone rock attacked by weak soil acids forms soluble bicarbonates as in the following equation.

$$CaCO_3 \ + \ H_2CO_3 \ \longrightarrow \ Ca(HCO_3)_2$$

Such activity by carbonic and other soil acids accounts for the rapid decomposition of calcareous rocks and eventual acidic reaction in soils developed upon them. Solution is, of course, most active in humid climates where dissolved materials are rapidly removed by water.

Oxidation of mineral matter often follows hydration as in this example using the mineral olivine:

$$3MgFeSiO_4 \ + \ 2H_2O \ \longrightarrow \ H_4Mg_3Si_2O_9 \ + \ SiO_2 \ + \ 3FeO$$

The ferrous oxide formed is oxidized to hematite.

$$4FeO \ + \ O_2 \ \longrightarrow \ 2Fe_2O_3$$

Iron and aluminum silicate minerals rapidly decompose under humid tropical conditions. Soluble bases of calcium, magnesium, potassium, and sodium are released from the mineral matter at a rapid rate and from rapidly oxidizing (decaying) organic matter as well, thus maintaining a pH (degree of soil acidity) close to the neutral point of 7. At pH7 the solubility of silica is increased and that of the ferromagnesian minerals (iron, aluminum and magnesium) is lessened, altering the character and color of the parent material. Oxidation and hydration continue to considerable depths, if the soil is at all permeable, resulting in a B horizon of great depth. The soil thus developed is the *latosol*; the process of development, *latosolization* (laterization).

If, on the other hand, decomposition of accumulated organic matter (*duff* or *mor*) occurs where there is high precipitation and low temperature, a distinct gray sub-horizon is formed in the A horizon.

The duff (accumulating usually from trees such as conifers and others where the base content of the leaves is low) will decompose under highly acid conditions. The acid leachate dissolves and removes much of the basic material from the soil, and the A horizon (having lost its bases) becomes highly acid with a pH of around 4.0. Iron and aluminum sesquioxides are removed, resulting in highly acid, silicious, gray (or ash-colored) A_2 horizon.

Colloidal organic matter is removed by the percolating water of relatively low pH and is deposited at the top of the B horizon as a dark layer of humus. This is typically underlain by illuviated sesquioxides or the sesquioxides may be carried downward to the water table. Thus original parent materials have been severely altered by the *podzolization* process. The soil profile exhibiting the characteristics of podzolization is the *podzol*.

Figure 2.8 schematically represents the calcification, laterization and podzolization processes with respect to climates. The tropics are represented by the Hot-Wet lower right hand corner; the high latitude, humid, cold climates are shown in the upper right hand corner as Cold-Humid; the dry continental interior, representatives of which are found on all continents, are represented by the Dry-Hot and Dry-Cold left-hand corners of the diagrams. The actual climatic zones are shown in Figure 1.1 in Chapter 1. As is shown, laterization is dominant in the wet tropics and decreases poleward as well as toward the dry continental interiors. Podzolization is most pronounced in the cool/wet middle to high latitudes and decreases in those regions of warmer and drier climates. Calcification (accumulation of calcium in the soil profile) is most dominant in the dry continental interior (desert and semiarid) areas of the world. In humid areas soluble salts may be removed in groundwater rather than accumulate in the soil. Figure 2.9 is illustrative of altitudinal changes (climatic changes) affecting soil development. Calcification is dominant in the desert soils and podzolization on the cold mountain slopes and peaks.

vegetation and biotic factors

Throughout earth history evidence exists of modifications in vegetation as modifications in the climate occurred. There is evidence, too,

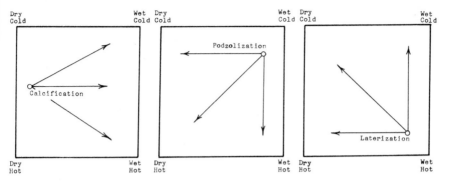

Figure 2.8. The processes of Laterization, Calcification and Podzolization are affected by climatic conditions. Laterization is most effective in the hot wet tropics and decreases toward the high latitude cold climates whether wet or dry. Calcification is most dominant in the dry continental interiors and Podzolization is most dominant in cold, wet, intermediate to high latitudes. The first and the last occur largely, although not exclusively, under forest cover as well.

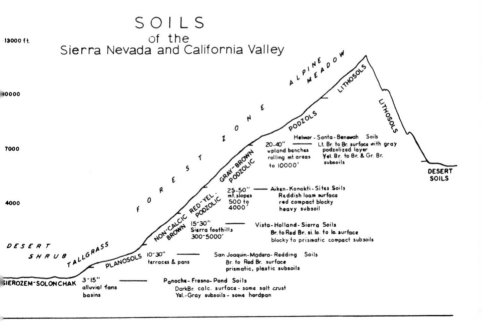

Figure 2.9. Soils vary with altitude as well as with climatic change. From the Sierozems of the California Valley up the west slope of the Sierra Nevadas to the Alpine meadows, the representatives of the Great Groups of soils exist in much the same pattern as they do in the latitudinal distribution of climatic regions. (Compare with Figure 2.11.)

of changes in the buried soils that formed under entirely different conditions than those existing on the surface today.

A soil often shows evidence of the plant and animal life growing upon it and active within it. The soil in equilibrium with its environment reflects the modifications caused by vegetation and organisms, as well as climate, topography and parent material. Most soils contain plant and animal organisms adapted to the existing soil content of air, water, oxygen, and carbon dioxide. Over many years a *climax vegetation* develops and adapts to the physical conditions of its particular area of the earth. In turn the climax vegetation affects each area in some way.

In a broad context the terms *forest soil* and *grassland soil* imply that there are certain characteristics of soils under a forest climax that are different from those of the grassland climax. The profiles will show these differences. Just as climate may limit the variety of vegetation that can exist within an area, vegetation in turn affects soil development. The soils of the mixed forest of the humid eastern United States are without question very different from those of the prairie and steppe grasslands of the Great Plains, whose climates are progressively more dry, the parent materials and topography notwithstanding. Figure 2.10 represents generalized profiles of two areas, the humid east and the subhumid interior of the continent. With the climate change from wet forested areas to dry shortgrass steppes the vegetation becomes less luxuriant and physically more widely spaced; the soil profile becomes less deep and organic matter less abundant; salts may accumulate throughout the profile in the drier areas and a lime zone is evident. The organic matter content of the surface soil reaches a maximum in the tall grasslands and decreases in all directions from that center.

Forest litter will break the impact of rain and allow for the slow penetration of rainwater through the litter and into the soil. It retards surface runoff and erosion. It forms an evaporation-retarding mulch on the soil surface in dry weather and a heat-retarding surface in cold weather and reduces frost action as well. It harbors microscopic animal life as well as larger forms, all aiding in the decay of the leaf litter into its simpler components. The burrowing and mixing actions of the various soil animals add fresh minerals to the soil, kill some roots, hasten the process of humus formation, and aid in the maintenance of good soil structure.

The forest soil is usually more open and porous than an agricultural soil under cultivation. It is more absorptive and therefore receptive to percolating water after sudden rains. It aids in the maintenance of

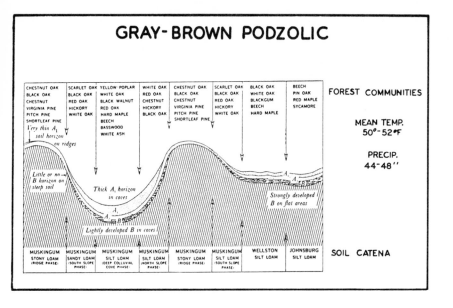

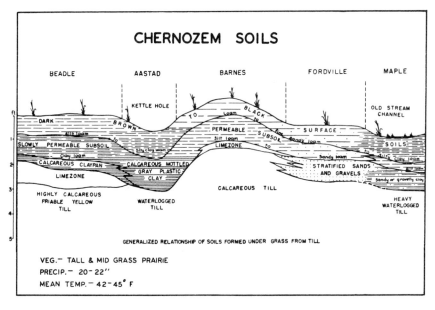

Figure 2.10. The upper Gray-Brown Podzolic profiles are representative of the non-glaciated, sharply dissected Appalachian Plateau where the precipitation is between 44 and 48 inches yearly. They vary considerably in thickness and stoniness as well. The Chernozem profiles, found in the western edge of the Central Lowlands in the Dakotas where the precipitation is half that of the east, show evidence of calcification with the developed lime zone. Upper illustration modified from U.S.D.A.

a higher water table and consequently in the reduction of flood hazards on the river bottoms because of its great capacity to hold rain water.

Breakdown and decay of the organic materials of the forest litter supplies the soil with humus. This highly decomposed organic matter is greatly reduced in agricultural soils since cultivation causes rapid oxidation. It should therefore be replaced systematically.

The trees of the deciduous forest absorb calcium, potassium, magnesium and the other mineral elements required for growth from the soil, and by means of decayed leaf litter return these elements to the soil. Although some mineral elements are lost through percolating water, weathering of rock minerals may in part replenish the supply. In an undisturbed state an equilibrium is maintained in which the forest litter accumulates and decays at approximately the same rate; the bases are lost through percolation but replaced from the decaying organic matter and minerals of the parent material at about the same rate; and the groundwater supply is maintained at a uniform level.

Any soil characteristic that affects water movement through the soil, the depth of the soil, or its porosity, will affect the vegetation upon it. The wetter soils in the depressions and upon the river terraces and floodplains support an entirely different biota than the drier soils of the hilltops or mountain slopes. Such changes in forest type have been observed over a relatively few years, both naturally and where man was instrumental in changing the soil environment. Figure 2.11 relates precipitation and climax vegetation on the west slopes of the Sierra Nevada, a relatively undisturbed area. The wetter western slope receives as much as fifty inches of precipitation. The eastern slope and base of the Sierra is desert, receiving only about ten inches of precipitation. A great variety in forest types is noted. On comparison with Figure 2.9 the relation of soils to vegetation will be seen for this same area of the Sierra Nevada mountains.

Trees failed to develop in the Great Plains because of unfavorable surface conditions, severe climatic conditions, or certain acts of man, although extensions of eastern forests are found on the north and east facing slopes and in sheltered areas in the grassland borders. Grasses, however, are much more capable of withstanding the seasonal excess moisture and extended drought of the subhumid and semiarid climatic regions.

There is a working relationship, then, between vegetation and soils: each affects the other and the total environment so as to bring it all into a state of equilibrium. Where climate is effective in restricting

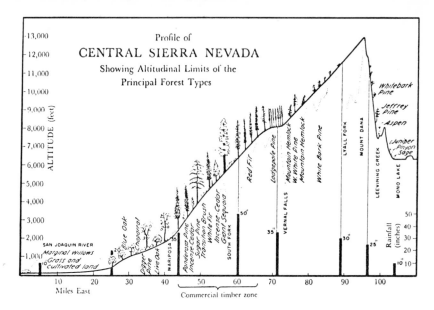

Figure 2.11. The Sierra Nevada west facing slope is considerably wetter than its east facing slope. The latter is in the lee of prevailing winds, considerably drier, and is in fact the western edge of the dry Basin and Range Physiographic Province in California. The altitudinal limits of the principal forest types are shown. Courtesy U.S.D.A.

germination of trees, grasses come in rapidly. Annuals often take over quickly on "made land," only to succumb to forest encroachment with passing time.

Soil color (partially a result of the organic matter content) is usually darker under grasses. Where sedimentary accumulations are received at a slow rate, newly deposited soil material may form soils at the same rate as weathering and decomposition occur. Also a rapid accumulation of humus and darkening of the surface soil layers will occur. However, where alluvium accumulates rapidly, the vegetation has little immediate effect upon the soil and its color may be largely that of the alluvium itself.

There is a further distinction made with relation to the organic matter found in the soil. The so-called *mineral soil* has upwards of 15 per cent to 20 per cent of organic matter in it, whereas the *organic soil* ranges up to 80 per cent or 90 per cent in some cases. The latter are not extensive, existing only in those wet environments favoring the accumulation of organic matter. Such bogs, marshes, and swamp areas are found in glaciated, coastal, and river valley areas. Under such conditions aquatic plants such as cattail, sedges, reeds, and mosses contribute to the bog or swamp large quantities of organic matter annually. Oxidation is retarded under water, thus preserving much of the organic matter in thick deposits. What decomposition of the organic matter that does occur is accomplished by fungi, anerobic bacteria, algae, and microscopic aquatic animals which give off gasses and form humus in the process. When the accumulations of organic matter

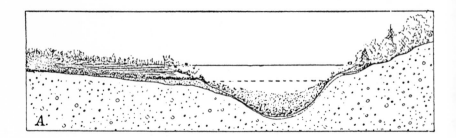

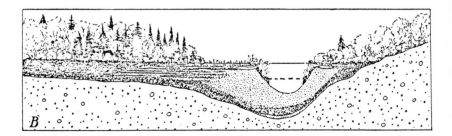

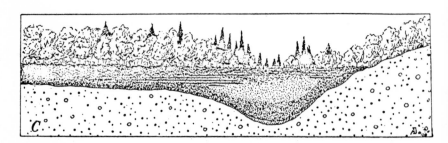

Figure 2.12. In the formation of an organic soil a shallow lake may be gradually filled with aquatic vegetation upon which marsh vegetation will develop and grow. Eventually the swamp forest in succession becomes the climax vegetation as shown in the three diagrams. Courtesy U.S.D.A.

become thick and firm and dry enough, the swamp forest is able to exist (Figure 2.12).

Peat and *muck* are the common organic soils. Peat contains much fibrous organic matter, usually above 50 per cent. Muck is composed of much more highly decomposed organic matter, unrecognizable as such to the eye. By definition, muck contains less than 50 per cent fibrous matter.

Schematic representations of the close relationships between climate, vegetation and soils are shown in Figure 2.13. The climatic variables of precipitation and temperature are shown in the corners as Hot-Wet, Cold-Wet, and so forth. Climatic types, zonal soil groups, and vegetative formations are shown on the climate base which if placed on the climate map of North America would have its lower left hand corner in the Sonoran Desert and the upper left corner high in the Arctic on approximately the same meridian. Its lower right-hand corner would lie on the east coast of Central America, and the upper right-hand corner would be in central Greenland.

Although shown as "close fits," the diagrams must be treated as though the boundary lines are transitional zones and it should be

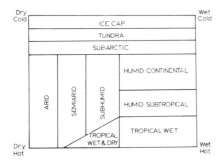

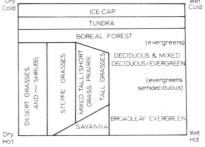

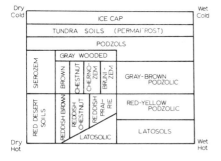

Figure 2.13. Schematically the relationships among climates, vegetation and soils may be shown as in this illustration. Boundary lines should be considered as transitional zones rather than sharp lines of demarcation. Were an attempt made to show the many micro controls operating singly or together, much more complex and probably incomprehensible diagrams would result.

understood that in comparing one with another only average conditions are depicted. Variations will occur in specific localities depending on the sum total effects of all control factors working together, or singly.

The very close relationship between vegetation and climate and between climate and soils has been shown by two studies that are of interest here as they, in a way, indicate the interrelationships between climate, vegetation, and soils. John R. Mather found a very good correlation between vegetation and climate by noting on a graph the vegetation type found at a station at a point, the coordinates of which are the potential evapotranspiration and moisture index of the station concerned.[3] *Potential evapotranspiration* is the amount of water that could be evaporated from the surface and transpired from plants. It is often greater than actual evapotranspiration, particularly early in the summer. The *moisture index* is a relationship between a computed water deficiency and water surplus within a year related to potential evapotranspiration (see Chapter 3 water balance discussion). In like manner, the author made a similar study showing good correlation between vegetation and soils in the north central United States. Figure 2.14 indicates the results of this study. Even though limited in its extent, the figure shows the general relationships found between potential evapotranspiration and moisture index of a station, and the climax vegetation and zonal soils of the same station for fifty different stations. The boundary lines are open at the top and bottom since the area studied was confined to the Central Lowlands between the Canadian boundary on the north and Kansas and Missouri on the south.

From the illustration one may note that short grasses are found in those areas where the moisture index is less than zero and the PE is between 22.5 and 31. Similarly, tall grasses range from a moisture index of −15 to +35 and a PE range of 22.5 to 31. There is no doubt that both would continue to higher as well as lower PE values if the area of study had been extended. Deciduous forests merge with the tall-grass prairies at a moisture index of +10 to +35; and insofar as the area studied is concerned the deciduous forests extend no farther poleward than to those stations where the PE is about 23. The mixed forest commencing at a PE of 23 extends to a PE value of about 21, at which point the Taiga is dominant. At these higher latitude stations,

---

[3]John R. Mather, and Gary A. Yoshioka, "The Role of Climate in the Distribution of Vegetation," *Annals,* Association of American Geographers, March, 1968, pp. 29-41.

the moisture index between tall-grass and forest decreases as it is between forest soils and grassland soils. A comparison of the distribution of zonal soils with respect to the moisture index and PE shows a good fit of zonal soils with dominant vegetation types above. Within the same general limits of PE and moisture index Chestnut and Brown soils are associated with short grasslands, Brunizems and Chernozems are associated with tall grasslands, and so forth. Better results should be obtained as more refined data becomes available. Since climate is not the most important factor in their formation, the two intrazonal soils (Hg and Pl) are not delineated and are left as they occur in association with the dominant zonal soil of the area.

Physical Properties of Soils

The physical properties of soils differ widely and affect the manner in which a soil may be used. The *pore space* (the openings between soil particles) may be large or small and as such affect the rate of movement of water through the soil and the amount of air that may be contained within it. After a short period of heavy rain, for example, all air may be excluded from the pore space as water is absorbed. As water moves downward, the rain having ceased, air replaces much of the water in the pore spaces. The amounts of organic matter, inorganic matter, water, and air within the soil vary widely from one climatic zone to another and from one season to the next, as well as from day to day. The average amounts of each were shown in Figure 2.3.

plasticity, cohesion and adsorbing power

The inorganic materials accumulating from rock weathering vary widely in size. These *soil separates* have size ranges, as shown in Table 2.1. Particles greater than 2.0 mm across are considered gravel, are inactive, and are not included as a soil separate, although when found in a soil mention is made of it in the profile description. The pore spaces in such cases are large and enhance movement of both air and percolating water, thus promoting good drainage and aeration.

The clay separate is smallest in size, being below 0.002 mm in diameter. Clay expands when wet, is very plastic and sticky and has a high water retention capacity, quite the opposite of that of the gravels mentioned above. Upon drying, shrinking occurs and large cracks develop in clay soils. This shrinking and swelling involves certain amounts of energy. The finer the particle size the greater is its

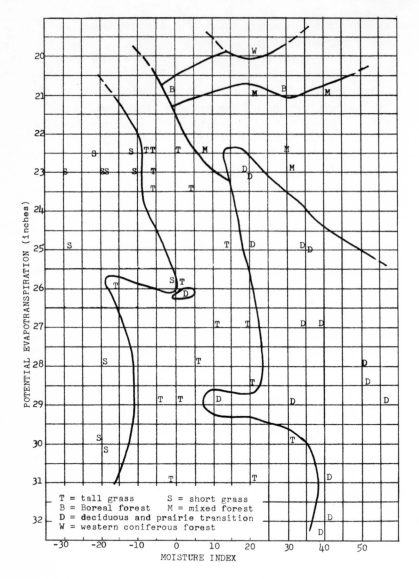

Figure 2.14. If climax vegetation and zonal soils on the above diagrams are compared within the parameters of the moisture index and potential evapotranspiration, some interesting relationships will be noted. The Chestnut and Brown soils (usually associated with the short grasslands) are found within the same general limits of potential evapotraspiration and moisture index as found for short grasslands. Both the Chernozems and the Brunizems are associated with tall grasslands, the Brunizems found on the more moist (higher moisture index)

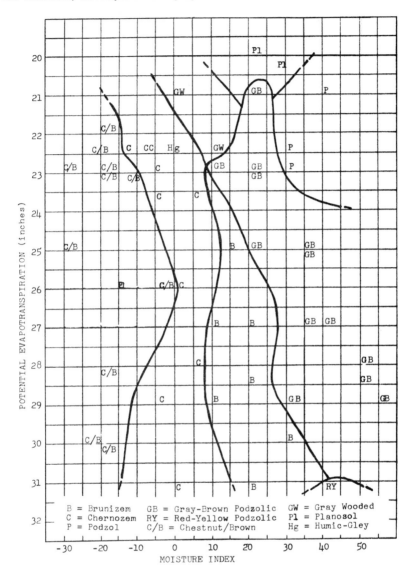

margin and the Chernozems on the drier margin. The Chernozems having a lime zone and the Brunizems having none, should mean that calcification process begins (or ends, since the Chestnut and Brown soils and others in the still drier areas also contain lime zones) at a moisture index of about zero. The other vegetation types and zonal soils may be correlated in a similar manner. Some Planosols found in association are not delineated.

TABLE 2.1

The classification of the soil separates as to their size limitations in the two major systems in use.

Soil Separate	U.S.D.A. System Diameter Limits (mm)	International System Diameter Limits (mm)
Very coarse sand	2.00–1.00	
Coarse sand	1.00–0.50	2.00–0.20
Medium sand	0.50–0.25	
Fine sand	0.25–0.10	0.20–0.02
Very fine sand	0.10–0.05	
Silt	.05–0.002	0.02–0.002
Clay	below 0.002	below 0.002

From *Soil Survey Manual* (U.S. Dept. of Agriculture Handbook No. 18, 1951), p. 207.

adsorption, plasticity, cohesion, swelling, and energy properties. Figure 2.15 shows in a relative way the almost negligible activity in the sands and silts and the tremendous adsorption, swelling, and plasticity in the finest of the clays. The clays are aluminum silicates and potassium aluminum silicates, and because of their activity in adsorption of other basic cations, supply much of the mineral nutrients to plants.

The *silt separate* ranges in size from 0.002 mm to 0.05 mm in diameter, is irregular in shape, and for the most part, composed of quartz. While silt is somewhat plastic and sticky, these physical properties are far less evident than in the clay separate. Although its chemical activity is less than that of clay, like clay, its capacity for holding water is high.

The *sand separate* particle size is considerably greater than that of either silt or clay, and its adhesive and plastic properties are negligible or entirely lacking. The coarser the sand the more easily water will percolate through it, as is the case with the gravels already mentioned. Both sand and silt are generally chemically inactive, supplying no cations to the soil solution, except for what may become available upon weathering.

Organic matter, particularly in a finely subdivided state, exhibits many of the physical properties of clay, although it is not as plastic.

As humus, it acts very much like colloidal clays in base exchange; in fact it has a considerably higher adsorptive ability than does clay itself.

soil color

The Munsell color charts are used to describe precisely what is probably the most readily noticeable physical property of a soil—its color. Although organic matter in various stages of decomposition and under different climatic regions imparts different degrees of brown and black color to the soil, humus is responsible for the blackness in many surface soils. Probably the blackest of these are the Chernozems of the semiarid regions of the midlatitudes. The deep red of the tropical

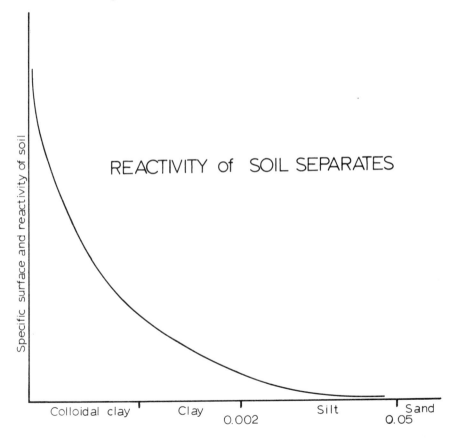

Figure 2.15. The amount of surface area affects the soil's chemical activity, it being greatest in the clay fraction. The swelling, plasticity and cohesion properties as well as the adsorbing power are greatest in the colloidal clays and decrease as the particle size increases.

laterized soils is unmistakable, as is the gray A₂ horizon of podzolized soils.

soil texture

The relative proportions of sand, silt, and clay found in a soil gives its textural classification. With specific amounts of each of the separates in a sample of soil, the physical and chemical properties of that soil will differ from another sample of different proportions. A *textural class* as shown in Figure 2.16 not only indicates what percentages of

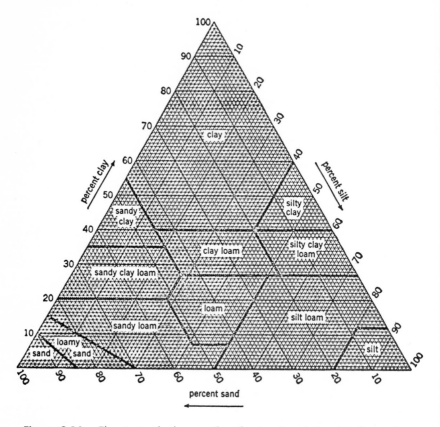

Figure 2.16. The textural classes of soils are found imprinted on the face of the nomogram. Three broad classes exist. A typical clay soil has in excess of 40% clay, a sand has 70% or more sand, and a silt soil has more than 80% silt. Combinations of the soil separates in different amounts will give different textural classes as noted. A typical loam, for example, is composed of about 40% sand, 40% silt, and 20% clay.

the various separates are in the sample, but from the content some indication of its physical properties may be deduced; the more clay, the more sticky and plastic with a greater base exchange capacity; the more sand, the more gritty and less sticky.

Three major textural classes exist. As recognized by the soil scientist, they are the *sands*, the *loams*, and the *clays*. The sands contain 70 per cent or more of material of particle size 0.05 mm or greater. To be a clay soil it must have about 40 per cent or more of clay separates. If the sample has considerable amounts of sand or silt it is referred to as a sandy clay or a silty clay, respectively. The loams cover a broader spectrum of particle sizes from 34 per cent to 40 per cent clay to about 70 per cent sand and 80 per cent silt. In this group the typical loam is of about equal proportions of sand and silt and around 10 per cent to 25 per cent clay. All of the textural classes are noted on the soil textural diagram and need not be repeated here. However, as was stated earlier, coarse sands and gravels may be present in some soils and this fact is noted in the textural classification of the soil in the field as coarse-sandy loam, gravelly sands, and the like.

soil structure

Whereas soil texture refers primarily to the size of soil separates, *soil structure* refers to the arrangement of soil separates into fairly distinct and usually larger secondary units or *aggregates* of fairly distinct shape and form. An aggregate is separated from adjacent aggregates by zones of weakness. The distinctive structural form is a result of certain biochemical and physical properties of the soil or parent material, as well as climate and other factors acting upon the soil.

Platy Structure

The aggregate mass is formed into thin horizontal plates or sheets each of which is an inch or so across and a very small fraction of an inch in thickness. The platy characteristic is frequently found in the eluviated horizon of the Gray-Brown Podzolic soils; however it may result from sedimentation of alluvial and lake plain sediments as well (Figure 2.17).

Columnar Structure

As the term implies, columnar structure is an arrangement of soil separates into vertically oriented aggregates, longer than they are wide. In the dry-land soils, a typical round-topped columnar structure, often salt encrusted, is commonly a result of degradation of the horizon.

Kind of structure	Description of aggregates (clusters)		Horizon
Crumb  🝆 ◔ ○	Aggregates are small, porous, and weakly held together	Nearly spherical, with many irregular surfaces	Usually found in surface soil or A horizon
Granular  ◖ ◑ ○	Aggregates are larger, harder, and strongly held together		
Platy 	Aggregates are flat or plate-like, with horizontal dimensions greater than the vertical. Plates overlap, usually causing slow permeability		Usually found in subsurface or A₂ horizon of timber and claypan soil
Angular blocky or cube-like 	Aggregates have sides at nearly right angles, tend to overlap	Nearly block-like, with 6 or more sides. All 3 dimensions about the same	
Subangular blocky or nut-like 	Aggregates have sides forming obtuse angles, corners are rounded. More permeable than blocky type		Usually found in subsoil or B horizon
Prismatic 	Without rounded caps	Prism-like with the vertical axis greater than the horizontal	
Columnar 	With rounded caps		
Structure lacking Single grain	Soil particles exist as individuals such as sand and do not form aggregates		Usually found in parent material or C horizon
Massive	Soil material clings together in large uniform masses, as in loess		

Figure 2.17. The arrangement of the soil separates is variable. The most common forms are as indicated in the figure. Courtesy of the University of Illinois, Cooperative Extension Service.

(See the soil profile of the Solenetz-solodized profile, Figure 4.9 where at about the nine-inch level this is observable.)

Blocky Structure

Where the aggregates of soil particles are roughly six-sided blocks or cubes, the dimensions of which are from a fraction of an inch to two or more inches, a blocky structure exists. Subtypes are recognized as *angular blocky* where boundary planes are sharply angular, and *subangular blocky* where boundary planes are mixed rounded and plane faces.

Granular Structure

If the aggregates are small and rounded and of a fraction of an inch in size the term granular is applied; if porous, the aggregate is of a *crumb* structure.

Mechanical weathering such as freezing and thawing, wetting and drying, root growth, the mixing action of burrowing animals, earthworm activity, and various actions of soil organisms which incorporate organic matter in the soil, aid in the development of a good soil structure. There are tillage practices performed by the farmer that also aid in maintenance of soil structure. The reverse may be true when organic matter is not returned to the soil in a crop rotation or if the soil is cultivated excessively; then soils lose their porous, loose characteristics and become more compacted and dense. Aeration and water percolation is impaired, soil microorganism activities decrease, and in general the soil loses much of its crop producing ability. With lesser returns per acre from a farm the monetary return, of course, is less. The maintenance of good soil structure by all means possible is of primary importance on all cultivated lands.

Not only is good soil structure of prime importance to the farmer but to the developer and home owner as well. Removal or burial of top soil and leveling of the land with heavy equipment destroys the soil structure, creating drainage and percolation problems and possibly erosion problems as well. Instead of the loose, crumb structure, compaction reduces the sponge-like characteristics of the surface soil to a very low percentage of its original capacity for holding water. Instead of soaking into the soil where it will aid in crop production and maintenance of groundwater supplies, water runs off as surface water and is lost, or "ponds" in the depressions.

Constant tillage without due regard to maintenance of structure makes the soil "heavy." Relatively, clay increases in a tilled soil when

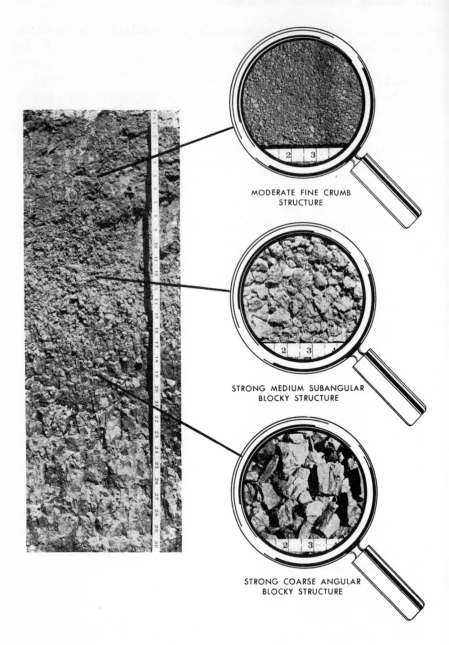

Figure 2.18. The structures found in this profile are indicated for the 4 inch, 12 inch and 20 inch levels. Courtesy of the University of Illinois, Cooperative Extension Service.

MODERATE FINE CRUMB
STRUCTURE

STRONG MEDIUM SUBANGULAR
BLOCKY STRUCTURE

STRONG COARSE ANGULAR
BLOCKY STRUCTURE

its B horizon is exposed by erosion, or when organic matter within the profile is destroyed. The soil is then more difficult to plow. On the other hand, where a crumb structure is maintained it is referred to as a *light* soil and is easier to plow. The terms *heavy soil* and *light soil* are in a sense textural-structural terms that refer more to the ease of working the soil than specific physical properties of the soil, although of course they are related. Figure 2.18 is a photograph of a profile with magnified views of the structure at three levels within the profile.

SELECTED REFERENCES

ARNOLD, R. W. "Multiple Working Hypothesis in Soil Genesis." Soil Science Society of America *Proceedings*, Nov.-Dec., 1965, pp. 717-724.

AUTEN, J. T. and PLAIN, T. B. "Forests & Soils," *Trees, USDA Yearbook of Agriculture*, 1949, Washington: 1949, pp. 114-119.

BIDWELL, O. W., and HOLE, F. D. "Man as a Factor in Soil Formation," *Soil Science* 99(1965):65-72.

BUNTLEY, G. J. and WESTIN, F. C. "A Comparative Study of Development Color in a Chestnut–Chernozem-Brunizen-Climosequence," Soil Science Society of America *Proceedings*, July-August, 1965, pp. 579-582.

FINNEY, H. R.; HOLOWAYCHUK, N.; and HEDDLESON, M. R. "The Influence of Microclimate on the Morphology of Certain Soils of the Allegheny Plateau of Ohio," Soil Science Society of America, *Proceedings*, May-June, 1962, pp. 287-292.

KUCHLER, AUGUST W. *Potential Natural Vegetation of the Conterminous United States*. American Geographic Society Special Publication, no. 36 New York: AGS 1964.

MATHER, JOHN R. and YOSHIOKA, GARY A. "The Role of Climate in the Distribution of Vegetation," *Annals*, Association of American Geographers, March, 1968, pp. 29-41.

SHANTZ, H. L. "The National Vegetation of the Great Plains Regions," *Annals*, Association of American Geographers, Vol. 13, 1923, pp. 81-107.

Soil Conservation Service. "A Toposequence of Soils in Tonalite Grus, in the Southern California Peninsular Range," Soil Survey Investigations, *Report* 21, USDA, Washington: 1968.

Stark County Regional Planning Commission. *Physical Geography–Stark County, Ohio–A Planning Dimension*. Canton, Ohio: 1960. Contains two very useful maps, predominant soil areas and selected characteristics and major physical barriers to development.

TEDROW, J. C. F. "Polar Desert Soils," Soil Science Society of America *Proceedings*, May-June, 1966, pp. 381-387.

THORP, JAMES. "How Soils Develop Under Grass," *Grass, USDA Yearbook of Agriculture*, 1948, Washington: 1948, pp. 55-66.

Chapter 3

Soil Temperature,
Soil Water, and Water Balance

Soil temperature is directly related to the receipt of energy from the sun. The intensity of solar energy (or radiation) may be reduced by absorption, reflection, and refraction in the atmosphere before reaching the earth surface. Annually about half is dissipated within the atmosphere and half absorbed by the earth. Solar radiation in any one latitude varies seasonally and is subject to further losses depending upon the character of the surface of the earth.

Some radiation is reflected from the soil, plant leaves, water bodies and other objects. The receipt of solar radiation may be affected adversely by the degree and orientation of slope and the color of the soil. That which is absorbed by the soil is converted to heat energy and thus affects soil temperature. Figure 3.1 indicates the average distribution of solar radiation within the atmosphere and at the earth surface. Thirty-four per cent passes through the atmosphere and reaches the earth directly and another 17 per cent comes from that scattered by clouds.

If the incident solar radiation is received on the surface at an angle, much is reflected. It is possible to have this solar energy strike the surface at such a low angle that total reflection may occur. Should this happen no energy is available to enter the soil. This reflectivity of the earth varies greatly with the seasonal position of the sun. The only places where the sun will be perpendicular at some time during the year are those geographic areas that lie between the Tropics. There are exceptions. The south facing slopes in the northern hemisphere may receive the direct rays of the sun; then the absorption at the surface is at its greatest. On north facing slopes in the northern hemisphere (vice-versa south of the Equator) the sun's rays always strike at a very low angle or do not strike the surface at all. Changing

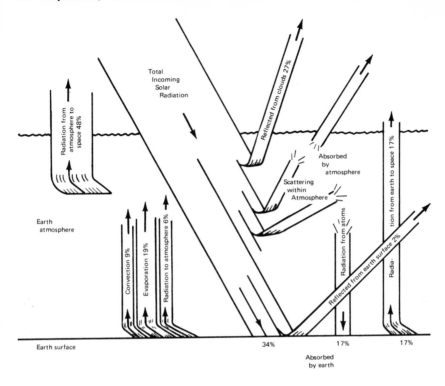

Figure 3.1. The incoming solar radiation is affected by the earth's atmosphere in a number of ways, so that only about half of the incident radiation is absorbed at the surface of the earth. Conduction, evaporation, and radiation from the earth's surface are ways that the energy is lost and soil temperatures lowered.

slope characteristics then, varying as they do, will affect the energy receipt at the surface from almost nothing with total reflection by the earth surface to near total absorption.

Soil color affects energy receipt, and hence soil temperature. The *albedo* or reflective power of a surface is greatest for the light-colored soil and least for the dark. A clean tilled black soil has a very low albedo; thus it is capable of absorbing most of the energy striking it and will be relatively warm. The sandy desert soil, having a very high albedo, will absorb very little of the solar energy and will be relatively cool.

Some ranges for the albedo of different surfaces are shown in Table 3.1. A comparison of the dry sand with the wet sand reveals that the

TABLE 3.1

Ranges for the albedo of different surfaces are given.

Albedos of Different Surfaces			
New snow	75-95	Deciduous forest	10-20
Old snow	40-70	Coniferous forest	5-15
Dry sand	35-40		
Wet sand	20-30		
Dark soil	5-15		
Gray soil	20-35		

albedo is lowered when the sand is wet. This will be generally true on all wet soils. The darker soils, if devoid of vegetation and oriented to receive the direct rays of the sun, should absorb practically all the energy reaching the surface of the earth at that point.

The energy required to raise the temperature of one gram of soil 1° C is the *specific heat* of the soil. It is different for different soils and will further vary if the soil is wet. Moisture in a soil affects the specific heat of the soil by requiring more heat per unit volume. Thus a wet soil will heat and cool more slowly than a dry soil. Consequently, sharp temperature changes are least likely in a wet soil. It is because of the slow conductivity of heat through and slow heating of a wet soil in the spring, that a wet soil is often referred to as a *cold* soil.

It takes energy to evaporate water and evaporation of water from the soil lowers the soil's temperature, a significant fact at planting time when germination of seed will be delayed in the cold soil.

While evaporation causes a loss of heat in soils, *conduction* is the means by which heat is carried downward and laterally through the soil. Here, too, the moisture content of the soil is important. Within soil pore spaces heat passes from soil to water much more readily than from soil to air. Therefore in the wet soil the transfer of heat progresses more easily. At night when solar radiation is no longer received at the soil surface, radiation from the soil surface to the cooler atmosphere takes place, cooling the surface. Diurnal (daily) temperature fluctuations are greatest on the soil surface and decrease with depth (Figure 3.2). Seasonal temperature fluctuations react similarly.

It takes time for the transfer of energy downward through the soil. By the time (and it may be many hours) the higher surface heat moves

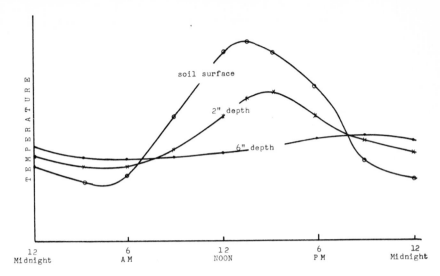

Figure 3.2. The soil being a poor conductor of heat concentrates much of the absorbed energy at its surface. The temperature fluctuations within a 24-hour period will be greatest on the surface and least at depth. Seasonal variations are very similar in that the shape of the curves would appear much as they do for the average day.

to lower levels, radiation from the surface at night causes a cooling of the surface, and a reversal of heat flow from the warmer, lower soil to the cooler surface soil follows. Because of a temperature lag, fluctuations are greatest at the surface and decrease with depth. Figure 3.3 indicates that greater seasonal ranges at the surface of a typical mineral soil disappear at depth, reaching almost isothermal conditions in the bedrock below. Seasonal variations in an organic soil, particularly if wet, would be considerably less than that shown for the mineral soil. Thus the fluctuations in soil temperature are related to the heat capacity, evaporation, and thermal conductivity of the soil and these in turn are affected by the amount of mineral matter, water and air that the soil contains.

Vegetation is an effective insulator of the surface soil from the sun radiation, and keeps soil temperatures from reaching high levels. Mulches have been used on agricultural soils to reduce wide fluctuations in soil temperature. The better mulches retard heat flow in and out of the soil, consequently the surface temperature is maintained at a more uniform level with minimum fluctuation between day and night. If graphed it would look much like Figure 3.2 with

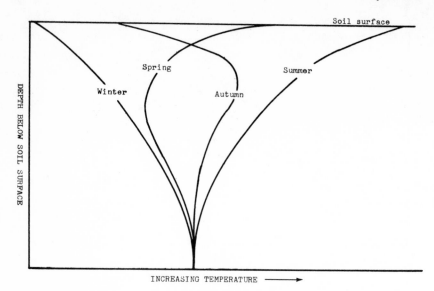

Figure 3.3. The wide range in surface temperature between summer and winter is evident. Subsurface temperatures vary less widely and become almost isothermal in the bedrock below.

all curves flattened. Since soil temperatures are so closely related to soil water content any manipulation of either will affect the other.

Soil and Water Balance

In dry climatic areas various procedures to reduce evaporation from the earth surface have been used. They include covering the soil with a mulch as mentioned, or plastic films, ponds and reservoirs with a chemical film, and the use of plants that require lesser amounts of water for their life processes, the latter to reduce transpiration losses.

Some soil water moves quickly through pore spaces to the water table and is essentially unused in the soil. This is *free* water. *Capillary* water moves by capillarity, is evaporated from the soil surface or is taken up by plants. *Hygroscopic* water is held tightly by the colloidal complex and is unavailable for any purpose.

In an effort to determine the soil water a number of empirical means have been devised to indicate the amount available at any one time—a soil water balance sheet showing water surplus and deficit in the soils of the area. A comparison of the amount of precipitation received, with the measured losses through evaporation, transpiration and runoff, will give some indication of the amount of water made available in the soil. This procedure requires special equipment. However, it is used as a check for most empirical means of determining evapotranspiration.

One procedure that has been employed to determine the soil water availability uses a graphic means for showing water surplus and water deficiency. It is explained briefly below.

The amount of solar radiation and the latitudinal position of the station are taken into account. By an adjustment procedure for latitude and seasonal solar radiation receipts, weather records of temperature and precipitation data are utilized to determine the *potential evapotranspiration*. When the temperatures are high, evaporation and transpiration are high, higher possibly than the water supplied as rainfall. The annual potential evapotranspiration is much higher than precipitation in desert regions. This is true to a lesser extent in humid regions; nevertheless, potential evapotranspiration is frequently higher than the precipitation in humid regions some months of the year (with the possible exception of the tropical rainy climates).

By this particular soil water balance technique the potential evapotranspiration is computed and plotted against the precipitation for each month. Two intersecting curves result. The potential evapotranspiration curve will be higher than the precipitation curve in the spring and summer for most stations.

Depending upon texture, all soils have the capability of storing varying amounts of water, and when precipitation is not available, supplying it to plants to cover their water needs. Should it rain the soil water removed may be in part if not entirely replenished. Such a month by month water balance computation (or even a day by day computation) has been utilized by the irrigator. From it he may obtain a very good idea of the amounts of irrigation water he will need at any time. Using a relatively simple bookkeeping procedure, the additions and depletions of soil water are made in the ledger each month to show the water need, water deficit, and water surplus, as well as other variables for the area.

Such a graph is Figure 3.4, developed for Columbus, Ohio. On it, water surplus, water deficit, water utilization from the soil, and recharge of soil water are shown. Similar computations of water surplus and water deficiency may be obtained for each of many stations statewide and nationwide. This was done for about 60 stations in the state of Ohio and mapped. In this case a computer mapping technique produced the maps (Figure 3.5 and Figure 3.6).

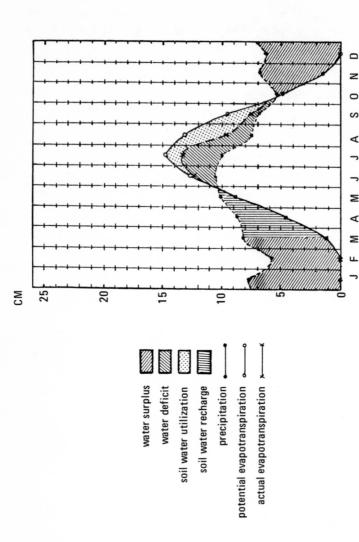

Figure 3.4. Potential evapotranspiration and actual precipitation are utilized to obtain water deficit and water surplus conditions for a station. The PE is calculated. The water balance for Columbus, Ohio, using the Thornthwaite procedure is shown on the graph. Soil texture is a factor in determining the quantity of water that the soil may store.

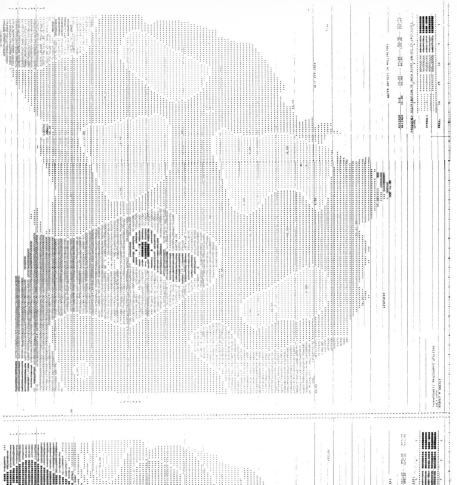

Figure 3.6

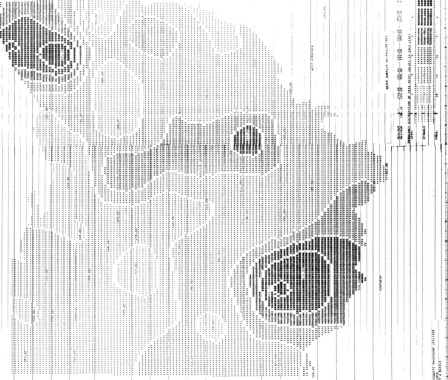

Figure 3.5

There should be no question but that the soil's ability to hold and release water on demand is an important soil property, particularly critical in those areas where the water supply is low. The ability of a soil to hold water can be increased by good management practices, thus in effect, increasing soil water supplies.

SELECTED REFERENCES

BUCKMAN, HARRY O., and BRADY, NYLE C. *The Nature and Properties of Soils*, 7th Ed., New York: The Macmillan Co., 1969.

CARLSON, J. E. and MOSES, H. "The Annual and Diurnal Heat-Exchange Cycles in Upper Layers of Soil," *Journal of Applied Meteorology*, June, 1963, pp. 397-406.

GARY, HOWARD. "Some Temperatures under Forest and Grassland Cover Types in Northern New Mexico," *U.S.D.A. Forest Service Research Note RM 118*, 1968.

JENSEN, MARVIN E. "Empirical Methods of Estimating or Predicting Evapotranspiration Using Radiation," *Evaporation and it's Role in Water Resources Management*, Conference Proceedings, December, 1966, American Society of Agricultural Engineers, 1966.

THORNTHWAITE, C. W. and MATHER, J. R. "The Water Balance," *Publications in Climatology*, Vol. VIII, no. 1, Centerton, N.J.: 1955.

TAYLOR, JAMES A., ed. *Aspects of Soil Climate*, Synopsis of Discussions, March 8, 1961, University College of Wales, Aberystwyth: 1961.

"Soils and Men," *USDA Yearbook of Agriculture*, 1938, Washington: 1938.

"Soil," *USDA Yearbook of Agriculture*, 1957, Washington: 1957.

World Distribution of Soils

The world distribution of soils as shown in Figure 1.1 closely follows the distribution patterns of climates (Figure 1.3) and vegetation (Figure 1.2). A more detailed, large scale map of soils will, of course, show many variations, reflecting topographic position and slope characteristics (north versus south facing slopes, for example, which under many circumstances will have a climate all their own) and parent materials. On a large scale map the effects of each of these is often recognizable as soil boundaries become more intricate and detailed. A comparison of the world, U.S. (Figure 4.1), state, (Figure 4.2), and county[1] soil maps will indicate the increased detail available. In fact, a large scale soil map to a considerable extent indicates the complexity of the natural environment. The soil profile synthesizes all these factors, as well as time, *in situ*. It is the "combined expression" of all these forces working together. And since the possible combinations of climatic, biological, and time of development are legion, the number of soil types is very great. The larger the map scale used, the more detailed will be the complex of soils shown and information obtainable.

It was a Russian soil scientist, Dokuchaiev, who about 1870 recognized the soil to be as much a dynamic and independent part of the landscape as the flora or the fauna or the rock structure. The soil was recognized as having a definite morphology of its own resulting from a unique combination of climate, vegetation, parent materials, topography and age, and not simply the weathered lithic materials of the earth's crust. Since the soil was recognized as an independent natural body, environmental factors were considered unimportant and

---

[1]The Soil Conservation Service has an index of published soil surveys listing by state all the county soil surveys available.

a classification based upon soil profile characteristics themselves was attempted. To the trained soil scientist the combined effects of all of the factors in soil formation is the soil profile itself.

Dokuchaiev and his students classified the soils of Russia using this genetic concept of soil development.[2] They recognized that the unconsolidated earth material was not a soil, but rather a medium in which a soil evolved. In the Russian soil classification scheme the three major groups (orders) of soils were recognized: the zonal soils, intrazonal soils, and azonal soils. The *zonal* soils are those with well

---

[2]"Soils and Men," *USDA Yearbook of Agriculture, 1938*, p. 980.

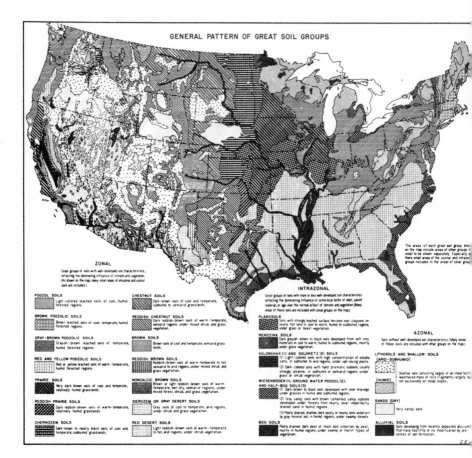

Figure 4.1. Soils map of the U.S. classified according to the system in use until the 1950's.

OHIO SOIL REGIONS

PUBLISHED BY
Ohio Department of Natural Resources
Division of Lands and Soil
Columbus, Ohio

Printed   1954
Revised   1956
Revised   1962

Scale in Miles

THE OLD LAKE-BED REGION - dominantly dark colored poorly drained soils formed from clay.

HIGH-LIME WISCONSIN TILL SOIL REGION - a gently undulating to rolling till plain.

ILLINOIAN HIGH-LIME TILL SOIL REGION - mainly old deeply weathered soils.

LIMESTONE AND SHALE RESIDUAL SOIL REGION - soils are low to high in fertility on undulating stream dissected topography.

OLD LAKE-BED REGION OF NORTHEAST OHIO - soils are lower in fertility and more acid than in Northwest Lake-Bed Region.

LOW-LIME TILL REGION - dominantly acid soils on undulating to rolling topography.

SANDSTONE AND SHALE RESIDUAL SOIL REGION - shallow to moderately deep acid soils on strongly sloping to steep topography. Thin-bedded limestone and limey shale occur in some of the eastern parts of the region.

Figure 4.2. The detail of soil subdivisions shown here is much greater than that shown on the soils map of the United States. Soil associations are shown only by a number within irregular boundary lines on the map. Each number represents two or three soil series.

developed soil horizon characteristics reflecting the influence of the primary factors of soil formation, climate and vegetation, and to some extent implying latitudinal zonation of the earth surface. The *intrazonal* subdivision includes those soils with fairly well-developed profile characteristics but reflecting primarily the strong effect of some local environmental factor of relief, parent material or age, and secondarily or not at all, climatic and vegetation controls. The third, *azonal* soils, are those which have poorly developed profiles often the result of recent deposition of alluvium or some quality of parent material or relief that inhibits profile development. Under these were thirteen subclasses identified by names of which "Chernozem" and "Laterite" are probably familiar, "Skeletal" soils and "Solonetz" soils much less familiar.

As developed in the United States, soil classification evolved from an early emphasis upon environmental features to a later emphasis on geologic origin of parent materials. The latter reflected the interest and work of the geologists rather than composition, textural, and structural characteristics of the soils themselves. Eventually this emphasis on geologic origin resulted in one kind of parent material becoming a single soil series. Under this system Miami silt loam, Miami sandy loam, Miami loam, and all other Miami soil types as we know them today were simply listed as the Miami soil series developed on glacial till. All the variations in texture and subtleties of profile were not recognized, and a classification on the basis of the distinct soil profile characteristics as later developed was nonexistent. In time this was rectified largely through the efforts of C. F. Marbut (then chief of the U.S. Soil Survey) whose work led to the evolution of a more meaningful classification of soils. Each soil series as heretofore used was broken up into smaller units, based on textural characteristics of the soil surface. This unit, the *soil type,* is the basic unit in soil classification. Marbut called this Category I. In all, he grouped soils into six categories, the broadest of which is Category VI, the most elementary, Category I, (Table 4.1).

Category VI, or *soil order,* consists of zonal, intrazonal, and azonal soils as defined elsewhere. Each of these is further subdivided into *pedocals* and *pedalfers.* The *pedocals* were defined as those soils in which the accumulation of carbonates of calcium or calcium and magnesium is found throughout the profile; and *pedalfers,* soils in which iron and aluminum compounds are found and in which calcium carbonates are lacking. The terms were coined from *pedo* (Greek for ground) and *calx* (Latin for lime) and *alumen* and *ferrum* (Latin for

TABLE 4.1

Soil categories of the old classification system.

Category VI—	Pedalfers (VI-1)	Pedocals (VI-2)
Category V—	Soils from mechanically comminuted materials. Soils from siallitic decomposition products. Soils from allitic decomposition products.	Soils from mechanically comminuted materials.
Category IV—	Tundra. Podzols. Gray-brown Podzolic soils. Red soils. Yellow soils. Prairie soils. Lateritic soils. Laterite soils.	Chernozems. Dark-brown soils. Brown soils. Gray soils. Pedocalic soils of Arctic and tropical regions.
Category III—	Groups of mature but related soil series. Swamp soils. Glei soils. Rendzinas. Alluvial soils. Immature soils on slopes. Salty soils. Alkali soils. Peat soils.	Groups of mature but related soil series. Swamp soils. Glei soils. Rendzinas. Alluvial soils. Immature soils on slopes. Salty soils. Alkali soils. Peat soils.
Category II—	Soil series.	Soil series.
Category I—	Soil units, or types.	Soil units, or types.

aluminum and iron respectively) and were recognized in the first case to be closely related to the arid and semiarid climate zones in the United States and to the humid climate areas in the second. Both terms are considered obsolete in modern soil science since they are not mutually exclusive as formerly supposed.

Lower categories of soils were described on the basis of their other characteristics. Category V, the *suborder,* consisted mainly of soils formed on mechanically fragmented rock, allitic material, decomposed rock materials in which silica has been removed and aluminum and iron compounds predominate, and siallitic soils in which basic materials have been removed and which consist mainly of weathered aluminum clay material. Both terms, allitic soils and siallitic soils, are considered obsolete today.

Category IV, the *Great Soil Groups,* is of most importance in a study of the geography of soils. The areal extent of the Great Soil

Groups, delineated in Figure 1.1, may be readily compared with other maps of climate, vegetation, topography, and parent materials. The terminology, although of long standing, is giving way to the new terminology of the *7th Approximation* used by the Soil Conservation Service and discussed later. Categories III, II, and I are successively finer and finer subdivisions. Categories I and II have geographical connotations as well.

In the field the color, texture, and structure of each horizon is noted, as well as the thickness, arrangement, and chemical composition of each, and the geologic stuff making up the parent material. The characteristics exhibited by the soil profile are basic to a soil type. In Category I, the *soil type*, it is the texture of the surface soil that sets it apart as the most elementary unit. In Category II, the *soil series*, a broader more inclusive grouping of soils of similar profile is made. In it, all characteristics of the profile except texture are the same. (In the soil series as envisaged by Marbut there can be many different textures and therefore many soil types, so long as the other characteristics are similar.) Ideally the soil series should have but one soil type, a goal classification specialists are attempting to reach. The name given to a soil series is taken from a place name within the geographic locality where it was first studied. The name also implies some length and breadth, and zonality to the series.

In Category III, *families* of soils, a still broader classification between series and the great soil groups is made. It includes one or more series as a taxonomic grouping in which natural environmental relationships are important considerations. Those included in a family are similar in most respects except that one series might be derived from glacial till and a second from closely associated stream terrace deposits within the till. For example, the Miami series is associated with the Crosby, Celina, and Brookston series all developed on calcareous glacial till. (Refer to Figure 2.7 block diagram of western Ohio soils). The Fox series, developed on stream terraces over gravels but with general profile characteristics similar to the Miami series, is included in the family since its profile characteristics are similar. Admittedly this has been a rather confusing concept and still is in the most recent attempt of classification of soils. The tendency today is to reduce the soil series to one textural classification, thereby eliminating the range of soil types composing a series. Thus, with one series equaling a textural class, the series will become, in time, the basic unit in the class.

The Great Soil Groups (whether zonal, azonal, or intrazonal) are fundamental to a study of soil geography and a part of the greater subject of physical geography. As latitude and altitude change, climates, associated vegetation, and soils change (at least zonal soils, although to some extent all do). Thus the study and mapping of distribution of climate and vegetation are parts of physical geography. The interrelationships are obvious. The distribution and morphology of landforms (another facet of physical geography) while not as important overall, is of sufficient importance in the development of some soil groups to set them apart within the same climate/vegetation zone. Thus, for example, the alluvial soils of the Mississippi or the Rendzina (Figure 4.10), and lithosols of Alabama, all lie within the otherwise Red-Yellow Podzolic soil group of the southeastern United States, an area of subtropical climate and forest cover. Generally not extensive enough to delineate on a small scale map, they do occur, however, and are scattered throughout and intermixed within the various zonal soil groups.

A classification of soils should not rely on what caused the soil profile to develop. Profile characteristics that can be observed, studied, and measured (both in the field and in the laboratory) offer the best means of differentiation among soils and the development of a good soils classification. Morphologic factors, not genetic factors, are stressed.

Zonal Soils

Of the thousands of soil types all may be placed somewhere within the approximately thirty great soil groups. These are really the most meaningful subdivisions for our purposes. And of these, the zonal soils are most extensive and significant. However, the intrazonal and azonal soils will not be slighted in the following discussion. (See Table 4.2).

lateritic soils

Lateritic Soils of the Forested Humid Subtropical and Tropical Wet and Dry climatic area. Consist of Laterite, Reddish-Brown Laterite, and Yellowish-Brown Laterite soils [Ultisols-Humults (Haplohumults, Tropohumults); Udults (Rhodudults)].[3]

Lateritic soils are very deeply weathered and "open" enough to have excessive drainage throughout the profile. The soil profile of the laterite exhibits a red-brown surface horizon, a dark red B horizon

[3]The statements in brackets here and below refer to Order-Suborder (Great Group) of the 7th Approximation discussed below and inserted here as a cross reference. Many subgroups that *should* be noted have been omitted at this point.

TABLE 4.2

A classification of Soils into Orders, Suborders and Great Soil Groups as modified.

Order	Suborder	Great Soil Group**
Zonal soils	1. Soils of the cold zone (Tundra Climate)	Tundra soils
	2. Light-colored soils of arid regions. (Steppe and Desert Climates)	Desert soils Red Desert soils Sierozem Brown soils Reddish Brown soils
	3. Dark-colored soils of semiarid, subhumid, and humid grasslands (Subhumid margin of Humid Continental, Humid Subtropical and Steppe Climates)	Chestnut soils Reddish Chestnut soils Chernozem soils Prairie soils (Brunizem) Reddish Prairie soils
	4. Soils of the forest-grassland transition (Dry margin of the Humid Subtropical; Dry Summer Subtropical Climates)	Degraded Chernozem Noncalcic Brown or Shantung Brown soils
	5. Light-colored podzolized soils of the timbered regions (Subartic, Humid Continental, Humid Subtropical Climates)	Podzol soils Gray Wooded, or Gray-Brown Podzolic soils* Brown Podzolic soils Gray-Brown Podzolic soils* Red-Yellow Podzolic soils
	6. Lateritic soils of forested warm temperate and Tropical regions (Tropical Rainforest, Tropical Wet and Dry Climates)	Reddish-Brown Lateritic soils* Yellowish-Brown Lateritic soils Laterite soils*
Intrazonal soils	1. Halomorphic (saline and alkali) soils of imperfectly drained arid regions and littoral deposits (Desert Climates)	Solonchak, or Saline soils Solonetz soils Soloth soils
	2. Hydromorphic soils of marshes, swamps, seep areas, and flats. (Wide range of Climatic types)	Humic Gley soils* (includes Wiesenboden) Alpine Meadow soils Bog soils Half-bog soils Low-Humic Gley soils* Planosols Ground-Water Podzol soils Ground-Water Laterite soils
	3. Calcimorphic soils (Humid temperate Climates)	Brown Forest soils (Braunerde) Rendzina soils
Azonal soils	1. (Found in practically all Climatic zones)	Lithosols Regosols (includes Dry Sands) Alluvial soils

*New or recently modified great soil groups
**Each Great Soil Group is further subdivided into many soil series and soil types.

over a red or "reticulately mottled" C horizon. The generally high temperatures and relatively high precipitation favor intense laterization. The silica is removed by leaching, causing a relative increase in the aluminum and iron oxide content and a decrease in the base-exchange capacity of the soil. The soils are commonly of low pH, porous, and excessively leached (Figure 4.3).

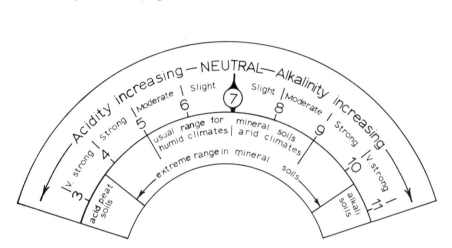

Figure 4.3. The pH scale for soils is given above. The extreme range of pH in mineral soils is from slightly under pH 4 to slightly above pH 10. Peat soils may have a pH which is less than that of mineral soils and the "alkali" soils of the arid and semiarid climates may run above pH 11.

Together with the Reddish-Brown and Yellowish-Brown Laterite soils, they have developed under a natural vegetation of broadleaf evergreen forest and/or savanna grasses and some semideciduous forest. The complexity of vegetation is great. However, little humus accumulation occurs as the high moisture and temperature are conducive to rapid decomposition by soil microorganisms.

Generally the soils are cropped as small farm units or occasionally as plantations of specialized crops such as pineapple and sugarcane. Although the soils are deep, they are likewise very porous. Rainfall percolates through them quickly, necessitating both irrigation and fertilization for their continued long-time agricultural use.

latosols

The *Latosols* of the Tropical Rainy climates [Oxisols-Humox, Orthox, Ustox; Ultisols-Humults (Tropohumults)] have developed in an environment in which temperatures and precipitation are always high. Soluble bases are released from the mineral matter with a subsequent pH of nearly 7. Although some Latosols have a great deal of organic matter in the surface horizons, this near neutral reaction is aided by the rapid decay of organic matter and release of bases from the organic materials. Silica under these conditions becomes soluble and the solubility of iron and aluminum is slowed. With continued weathering under high oxidation conditions, red and yellow parent materials are developed in which the silica sesquioxide ratio is low; i.e., silica is removed and the iron and aluminum sesquioxides accumulate.

The intense tropical weathering is instrumental in forming a very deep regolith which, regardless of the original lithic character, is fairly uniform. When exposed to the atmosphere the laterite horizon is generally soft and will harden irreversibly into a brick-like mass (Figure 4.4). The Latosol has a low base exchange capacity, a high degree of aggregate stability, and is usually red in color. Latosols, where cropped, are used for pineapple and sugarcane, sometimes under irrigation in the United States.

podzolic soils

Light-colored, podzolized soils of timbered regions consist of the Podzols, Gray Podzolic or Gray Wooded soils, Brown Podzolic soils, the Gray-Brown Podzolic soils, and the Red and Yellow Podzolic soils. Associated with the Boreal Forest (Taiga) of the Subarctic climates in high latitudes is the true *Podzol* [Spodisols-Orthods (Cryorthods, Fragiorthods, Haplorthods)]. In an undisturbed state the Podzols contain a few inches of leaf litter (mor) and solid humus underlain by a highly leached whitish-gray (podzol) A_2 horizon. The B horizon, low in bases, is heavier textured through illuviation and may contain a fragipan horizon (ortstein layer) in which the cementing materials are iron oxides and organic matter. The solum is usually no more than two or three feet in depth and is on the acid side throughout with a pH in some horizons as low as 3.5. This is particularly true on the humus layer which harbors a biota largely of fungi and a few bacteria. These northern latitudes, where podzolization is most intense, are largely areas of coniferous forest. The forest litter decomposes slowly, forming organic acids which, when leached, are influential in removing all soluble salts such as the carbonates and sulfates from the colloidal complex. In time all iron and aluminum are removed from the A_2 horizon, leaving the highly acid silica materials behind. The sesquioxides of iron and aluminum are deposited in the B_2 horizon as formerly stated.

Figure 4.4. The laterite crust is exposed in a road cut. The crudely blocky subangular masses are six to ten inches across. Courtesy N. Holowaychuk.

Productivity of the podzol is low for farm crops. However, under good management with the use of lime and fertilizers a relatively good tilth can be maintained, suitable for those few food crops (such as potatoes, some truck crops, hay and pasture) that will tolerate the severe climate.

The *Brown Podzolic soils* [Inceptisols-Andepts (Cryandepts); Spo-dosols-Orthods (Fragiorthods, Haplorthods)] are found in the hills and valleys south of the Podzol soils of New England. Not well developed, the Brown Podzolic soil seldom is more than two feet in depth and has a very thin gray leached A_2 horizon over a yellowish brown B in which there is some evidence of ortstein development. Developed on glacial till and outwash the Brown Podzolic soils in some instances tend to be stony or sandy. However, they are excellent soils on the stream terraces. Like the Podzol they are moderately to strongly acid and need liming and fertilization for good crop production. The climate is Humid Continental and not as severe as that of the Podzol area. The forests are mixed deciduous-coniferous types.

Middle latitude Podzolic soils are the *Gray-Brown Podzol soils*. [Alfisols-Udalfs (Fragiudalfs, Hapludalfs); Inceptisols-Andepts (Cryandepts); Ultisols-Humults (Hapludults)]. Developing under a Humid Continental climate they lie between the Podzol to the poleward and the Red-Yellow Podzolic on the equatorial margin. Toward the continental interior and on the less humid margin of the Humid Continental cimatic region, the Gray-Brown Podzols give way to the Brunizems. Developed under a climate of less severity than the Podzols, the Gray-Brown Podzols are somewhat less acid. The deciduous forest cover produces a thin leaf litter under which is a thin grayish brown eluviated zone to a depth of 8 or 10 inches. The B horizon accumulating colloidal material through illuviation, is heavier in texture, contains a high percentage of bases, and varies in color from yellowish brown to a light reddish brown. Soil acidity is in the medium range except for a slightly acid humus layer in a few cases. Where the parent materials are of glacial origin, as they are in much of midwestern United States, some intrazonal soils such as Planosols, Wiesenboden, and Bog soils may be found associated with them.

Those Gray-Brown Podzols developed on calcareous glacial till, loess, or from crystalline rocks (rather than from shales and sands) are the more productive (Figure 4.5). All are responsive to those management practices that improve the texture and structural characteristics and to liming and fertilization as well. Under such management production is high for a wide variety of crops including corn, soybeans and small grains. The hilly areas and those areas of thin soils are used for dairying and grazing.

The *Gray Wooded soils* [Alfisols-Boralfs (Cryoboralfs, Eutroboralfs, Fragiboralfs, Glossoboralfs)] are similar to the Gray-Brown Podzols in many respects yet are not as acid. They have developed in cool continental climates and at higher altitudes.

The lower middle latitude Podzolic soils are the *Red-Yellow Podzolic soils* [Ultisols-Udults (Hapludults, Paleudults, Fragiudults); Alfisols-Ustalfs (Paleudalfs)].

Under a warm climate with an abundance of moisture the Red-Yellow Podzolic soils have developed widely throughout the Humid Subtropical climatic zone of southeastern United States and other areas around the world of similar latitudes on the eastern sides of the continents. The soils are mildly to strongly acid in reaction. Yellow Podzolic soils are often found associated with a coniferous forest whereas the Red Podzolics are associated with a deciduous forest, although not mutually exclusive. The A horizon is strongly leached, yellowish gray to grayish brown, and varies in thickness up to several feet. The illuviated B horizon is red or yellow and often mottled. It contains a high percentage of illuviated and secondary clays and hence is more dense and heavy. Productivity is poor to medium but the soils will respond to fertilization and good management, and produce abundantly. Having been long used for row crop agriculture (cotton and tobacco among others), many of these soils have become exhausted and often do not exhibit distinct profile characteristics where plowed or where heavy surface erosion has occurred. Where abandoned, these lands revert quickly to forest, especially if aided by judicious plantings.

brunizem, chernozem and chestnut soils

The soils of the drier margins of the Humid Continental and Humid Subtropical climatic zones are the Prairie (Brunizem) soils, Chernozem and Chestnut soils (found in that order) in successively dryer areas eventually touching upon the Brown Steppe soils in the semiarid or Steppe climate. This is an area of the Great Plains within the United States where there have been great changes in the climate within a relatively short interval of time. The changes have been cyclical, from essentially a humid climate to an arid climate within historic times. The Brunizems are on the more humid side of this zone. The Chernozem and the Chestnut soils are on the semiarid side.

The *Brunizems* [Mollisols-Xerolls (Argixerolls, Durixeorolls, Palexerolls); Udolls (Argiudolls, Hapludolls)] have an A horizon that is dark brown to black and mildly acid, underlain by a B-horizon that is well oxidized and in which no zone of lime accumulation is found. The parent material, usually found from two to five feet below the surface, is variable in lime content (Figure 4.6). Brunizems have developed under tall grass prairies on the subhumid margin of the Humid Continental climate on glacial tills and loess for the most part. They are weakly podzolized, and are highly productive soils where well drained.

On the equatorward margin bordering on the Humid Subtropical climates the black color of the Brunizem changes to a dark brown or reddish brown A horizon and a reddish brown B horizon. These color characteristics are due in part to the color of the parent materials and to the greater oxidation of iron minerals in the warmer climate. These soils are the *Reddish Prairie soils* [Mollisols-Ustolls (Argiustols, Paleustols)]. Like the Brunizems the native grass cover each year through decay adds large quantities of organic matter to the surface horizons of the soils, thus aiding in the maintenance of a

Figure 4.5. Gray-Brown Podzolic soil formed from loess in eastern Iowa. These are fairly fertile soils although approximately 65% world-wide are potentially arable and as yet not fully used. Photo Soil Conservation Service.

Figure 4.6. A Brunizem soil on loess. The parent material starts at the lower arrow at about three feet. About 80% of the Brunizem soils of the world are potentially arable. Photo Soil Conservation Service.

granular structure and high exchangeable calcium on the colloidal complex.

The *Chernozem soils* [Mollisols-Borolls (Argiborolls, Cryoborolls, Haploborolls); Ustolls (Argiustolls, Haplustolls)] have a surface A horizon which is black or very dark grayish brown to a depth of three or four feet, grading into a lighter B horizon in which a zone of lime accumulation is found. This lime zone is one property the Chernozems possess that is distinctively different from those soils on its more humid borders.

Calcification, then, is dominant in the Chernozems soils (Figure 4.7). Precipitation, while adequate to maintain a native vegetation of tall and mixed tall-short grasses, is insufficient to leach the profile thoroughly. Consequently, an accumulation of lime, often concretion-like in form, is found at the lower solum, imparting a grayness to the otherwise yellowish brown parent materials. The pH in the subsoil is sightly alkaline at about 7.5 while at the surface it is slightly acid.

The Chernozems are located on the wetter margins of the semiarid or Steppe climates and adjacent to the Brunizems. The precipitation ranges around 18 to 28 inches and is unreliable. Because of this uncertainty of rainfall the Chernozems are of medium productivity. They will produce very well, however, if irrigated.

Degraded Chernozems [Alfisols-Boralfs (Eutroboralfs); Mollisols-Borolls (Haploborolls)] are found where forests have been encroaching upon the grasslands. A grayish A_2 horizon may be found along with evidence of a former lime zone in the B horizon. In some instances the climate is more humid than in the Chernozem zone and there is also evidence of podzolization following calcification.

The *Chestnut and Reddish Chestnut soils* [Alfisols-Ustalfs (Haplustalfs, Paleustalfs); Mollisolls-Ustolls (Argiustolls, Haplustolls); Xerolls (Haploxerolls)] lie in the semiarid climates and extend from intermediate to relatively high latitudes. Receiving about 15 inches of precipitation each year, the Chestnut soils have a cover of mixed tall and short grasses indigenous to steppe lands. Because of the lesser quantities of organic matter incorporated within the profile, their color is brown to dark brown, not the black of the Chernozems. Nor are they as deep. Surface soils are friable and dark brown in color. The B horizon is a brown with prismatic structure in which the lime zone may be anywhere from 1½ to 4½ feet from the surface. The productivity of the Chestnut and Reddish Chestnut soils is medium; moisture is the limiting factor to their agricultural use. Wheat and

Figure 4.7. A Chernozem soil formed from glacial till in South Dakota. Lime concretions are in evidence at about 2½ feet. Photo Soil Conservation Service.

Figure 4.8. The depth to secondary carbonates is at about 12 inches and below in these Brown Soils. Together with the Chestnut and Reddish Brown soils some 890 million acres are potentially arable. These are soils of the semiarid climates as the depth to lime implies. Photo Soil Conservation Service.

small grains, sorghums, and cotton are among the crops grown. Where irrigation water is available other crops may be grown; otherwise the land is used for grazing purposes.

soils of the steppe climates

In these relatively dry climates the *Brown* and *Reddish Brown soils* are found. [Mollisols-Ustolls (Argiustolls, Haplustolls) Xerolls (Argixerolls, Haploxerolls); Alfisols-Ustalfs (Haplustalfs, Paleustalfs)]. On the driest margins of the Steppe climates where the native vegetation consists of short grasses, bunch grasses, and shrubs only small amounts of organic matter are produced and incorporated within the soil (Figure 4.8). As a consequence the brown A horizon grades at a foot or two into a light gray, calcareous B horizon in which a lime zone is evident. The Brown soils are to be found in the middle latitude steppe; the Reddish Brown soils are most evident in the subtropical steppes. Drainage is generally good and when irrigated productivity of the land is moderate to high. Where the Brown and Reddish Brown remain unirrigated their best use is for grazing lands.

the soils of the arid climates

The soils of the deserts are the *Sierozem* or *Gray Desert soils*, the *Desert soils*, and the *Red Desert soils*. [Aridisols-Argids (Durargids, Haplargids, Paleargids); Orthids (Camborthids)]. In the middle latitude desert with precipitation of 10 inches or less, vegetation tends to be largely bunch grass, shrubs, and annuals. Desert soils are light gray or light brownish gray and low in organic matter. Calcification is dominant. Subsoils are light colored and contain a great deal of limey material. Since rainfall is very light, the soils are not greatly leached and contain an abundance of the mineral nutrients in an available form for plants. Where desert soils are very salty, little will grow. If the drainage is good, crop production can be quite good on the better soils.

The Red Desert soils are products of the subtropical desert, differing from the desert soils above in that a reddish-brown surface horizon and a more compact reddish-brown or red heavy limey subsoil is formed. The whole profile is only a few inches in depth. The Sierozem soils have a pale gray surface grading into calcareous material at a depth of a foot or less. Used mainly as rangeland (offering very poor grazing to livestock) the Red Desert soils will produce cotton, grain sorghums and small grain where irrigation water is available.

the soils of polar climates

Of the high latitude Polar Climates, the Tundra is the only one that has an identifiable soil; however, *Tundra soils* [Inceptisols-Umbrepts (Cryumbrepts)] are thin. Where long cold winters and very short summers with temperatures under 50° F exist, soil development is very slow. Native vegetation of lichens, moss, flowering plants, and shrubs contribute to a dark brown, peaty surface horizon. A grayish, mottled B horizon is underlain by a permanently frozen subsoil or permafrost. As an agricultural soil it is of little use, although as grazing land it may have future possibilities for herbivores (reindeer, muskox) indigenous to the cold high latitude climates.

summary

The zonal soils develop under different climate and vegetation types. They differ from one another sufficiently that distinct profile characteristics are observable. The dark surface coloration, the stamp of organic matter, or the heavy texture and massive structure of the subsoil, a result of illuviation, most often is obvious. It may be the limey deposits of the subsoil that are most revealing as to the soil's development, or it may be the columnar structure and high salt content throughout the profile. Or it may be nothing more than the red color of oxidized iron compounds that is most striking. Whatever the profile examined, certain characteristics will stand out to aid in the identification of the soil.

Intrazonal Soils

halomorphic soils

The intrazonal soils include the Halomorphic, Hydromorphic, and Calcimorphic suborders. Halomorphic soils, which have developed under poor drainage conditions and often are poorly drained, are found in the semiarid to arid climatic zones where 12 to 15 inches of precipitation would be considered wet. *Solonchak* [Aridisols-Orthids (Salorthids)] soils contain much light colored soluble salts of sodium, calcium, magnesium, and potassium, and are low in organic matter. Salinazation may produce salt crusts and where such salt crusts exist the land has a limited use for pastureland. In those areas where the Solonchak is better drained fruits and vegetables of great variety as well as other crops are grown under irrigation.

Where removal of exchangeable calcium and magnesium has been accomplished through drainage, sodium salt is adsorbed on the colloidal complex, and the Solonetz soils are formed. They are often found in low spots within Solonchak areas (Figure 4.9). In the *Solonetz* [(Aridisols-Argids (Naturargids, Natrargids); Mollisols-Ustolls (Natrustolls)] organic and mineral colloids are dispersed as a result of sodium saturation and are moved downward through the profile. The organic fraction accounts for the dark surface horizon. Often referred to as "black alkali soils," the Solonetz has a thin A horizon under which a hard clay B horizon of columnar structure (an easily identifiable characteristic of the Solonetz soil) is found. This is at about 9 inches depth on the photograph.

hydromorphic soils
The Hydromorphic soils are wet soils, often waterlogged, or so very slowly permeable in the subsoil as to have poor internal drainage. The Humic Gley soils which include the Wiesenboden or Meadow soils, Alpine Meadow soils, Bog soils, Half-Bog soils, Low Humic Gley soils, Planosols, Ground-Water Podzols, and Ground-Water Laterites are of this group.

The *Humic Gley soils* [Mollisols-Aquolls (Argiaquolls, Cryaquolls, Haplaquolls); Inceptisols-Aquepts (Humaquepts); Utisols-Aquults (Umbraquults)] are found primarily in association with the zonal Brunizem soils and are much like them when drained. Topographically they occur in poorly drained flat areas. The A horizon is dark brown to black and high in organic matter. The subsoil is gray rust, mottled between one and two feet. They are subject to *gleyzation*, a process which aids in the formation of a blue-gray clay B horizon. During the dry season some of the Humic Gley soils are dry part of the time. When drained they are very highly productive soils and are in fact among the best soils of the corn belt.

Alpine Meadow soils [Spodisols-Aquods (Cryaquods); Mollisols-Aquolls (Cryaquolls)] develop at high altitudes in a cold climate under grasses, sedges, and flowering plants. The profile is dark brown to black to about one or two feet, at which level it is grayish to white and mottled over a horizon of clay accumulation, and poorly drained. Their best use is for seasonal grazing lands.

Bog soils [Histosols-Fibrists, Hemists, and Saprists] are developed over considerable latitude under generally humid conditions. They are wet organic soils of swamps and marshes. They have a brown, dark brown or black peat or muck over a brown peat indicative of a cover of swamp forest sedges and grasses. As with all Hydromorphic soils they are poorly drained. When artificially drained they are productive for special crops.

The *Half-Bog soils* [Inceptisols-Histic subgroups of the Aquepts] are similar in most respects to the Bog soils with one major exception—they contain a grayish and rust mottled gley subsurface horizon.

Planosols [Alfisols-Aqualfs (Albaqualfs, Fragiaqualfs, Glossaaqualfs); Mollisols-Albolls Argialbolls] are the nearly level, poorly drained upland soils of the prairies and forest regions in the midde to low latitudes. Some have organic rich surface horizons which are medium to high in bases. They have a compact or cemented claypan, a distinguishing characteristic, which is responsible for the poor internal drainage. Because of the great range of latitude in which the planosols are found the processes of podzolization, gleyzation, and laterization have been instrumental in their development.

Ground-Water Podzols [Spodisols-Aquods] although not as poorly drained as the Half-Bog and Bog soils, nevertheless have developed under a high water table. At two or three feet is a characteristic ortstein layer (B_2 horizon). Above the ortstein is an organic mat and thin acid humus layer over a whitish gray leached mineral horizon. Found under humid, cool to tropical climatic conditions, the Ground-Water podzol is generally of low productivity.

calcimorphic soils

Calcimorphic soils owe their characteristics to parent materials of high lime content. The Brown Forest soils and Rendzina soils belong in this suborder of the intrazonal soils.

The *Brown Forest soils* (Braunerde) [Inceptisols-Ochrepts (Eutrochrepts)] have a brown friable surface soil and a lighter colored B horizon from which mineral materials have been removed above the calcareous parent materials. In reality, since little illuviation occurs, the layering of horizons is ill-defined. They have developed under broadleaf forest where podzolization is very weak, if it exists at all. The Braunerde of Europe is more extensive and of greater importance than its equivalent in the North American continent.

The *Rendzina* [Mollisols-Rendols] are usually dark grayish brown to black granular soils developed upon soft calcareous parent materials under a grassland cover. Good examples of these are found in Texas and Alabama (Figure 4.10).

The Azonal Soils

Azonal soils have characteristics that reflect not climate or vegetation but the nature of the parent materials. They include lithosols,

Figure 4.9. The Solonetz soils have a thin A horizon over a B horizon of heavy clay of a columnar structure. At about nine inches the columnar structure is visible with a white coating of exchangeable sodium. This profile is near Dickinson, North Dakota. Photo Soil Conservation Service.

Figure 4.10. The Rendzina, an intrazonal soil, has developed upon soft, calcareous parent materials. This Puerto Rican Rendzina has marl at about 12 inches.

regosols, and alluvial soils. The lithosols are thin stony soils over stony parent material found under a wide range of climatic conditions. They are common on steep slopes and mountain tops, show little or no evidence of illuviation, and agriculturally are of little importance except as pastureland. The major uses for lithosols are seasonal grazing and permanent pasture.

Regosols [Entisols-Psamments (Cryopsamments, Quartzipsamments, Torripsamments, Udipsamments, Ustipsamments)] are very young soils developed upon deep unconsolidated materials of a non-stony character. They are found in areas of sand dunes, loess, and steeply sloping glacial deposits; consequently they are in the main excessively drained and have little profile development. Agriculturally they are used primarily for grazing. The Nebraska sand hills area is a good example of these.

The *Alluvial soils* [Entisols-Fluvents (Torrifluvents, Xerofluvents)] are found along the major rivers in all latitudes where recent alluvium has been deposited. They contain virtually no profile. Where properly drained, alluvial soils yield very well and, in fact, do support large populations in many regions of the world. An exception, however, is alluvial soil of the high latitudes which have little or no agricultural value today. It is doubtful that these will ever be of much value.

SELECTED REFERENCES

Agricultural Experiment Stations, North Central Region, and USDA *Soils of the North Central Region of the United States.* Madison: University of Wisconsin Press, 1960.

Agricultural Experiment Stations, Western States Land-Grant Universities and Colleges and Soil Conservation Service, *Soils of the Western United States.* Pullman: Washington State University Press, 1964.

Bibliography. A Geographical Bibliography for American College Libraries. Commission on College Geography, Publication No. 9, Association of American Geographers, Washington: 1970.

EYRE, S. R. *Vegetation and Soils, A World Picture.* London: Edward Arnold (Publisher) Ltd., 1963.

GIBSON, J. SULLIVAN and BATTEN, JAMES W. *Soils—Their Nature, Classes, Distribution, Uses and Care.* Montgomery: University of Alabama Press, 1970.

BUNTING, BRIAN T. *The Geography of Soil.* Chicago: Aldine Publishing Co., 1965.

GLINKA, K. D. *Treatise on Soil Science* (Translated from Russian). Jerusalem: National Science Foundation and U.S. Department of Agriculture, Israel Program for Scientific Translations, 1963.

GERASIMOV, INNOKENTII P. and GLASOVSKAYA, M. A. *Fundamentals of Soil Science and Soil Geography,* (Translated from Russian) Jerusalem: National Science Foundation and U.S. Department of Agriculture, Israel Program for Scientific Translations, 1965.

IVANOVA, E. I., ed. *Genesis and Classification of Semidesert Soils,* (Translated from the Russian). Jerusalem: National Scientific Foundation and U.S. Department of Agriculture, Israel Program for Scientific Translations, 1970.

————. *Soils of Eastern Siberia,* (Translated from the Russian). Jerusalem: National Science Foundation and U.S. Department of Agriculture, Israel Program for Scientific Translations, 1970.

KUCHLER, AUGUST W. *Potential Natural Vegetation of the Conterminious United States.* American Geographic Society Special Publication no. 36. New York: American Geographic Society, 1964.

JENNY, HANS H. *Factors of Soil Formatioin: A System of Quantitative Pedology.* New York: McGraw-Hill Book Co., 1941.

KUBIENA, WALTER L. *The Soils of Europe.* London: Thomas Murby, 1953.

LOBOVA, E. V. *Soils of the Desert Zone of the U.S.S.R.* (Translated from the Russian). Jerusalem: National Science Foundation and U.S. Department of Agriculture, Israel Program for Scientific Translations, 1967.

SHANTZ, H. L. "The Natural Vegetation of the Great Plains Regions," *Annals* of the Association American Geographers, Vol. 13, 1925, pp. 81-107.

SIMONSON, ROY W. "Soil Classification in the United States," *Science,* Sept. 28, 1962, pp. 1027-1034. A concise review of attempts at classification but more important is a discussian of the 7th Approximation.

Soil Survey Reports. The Soil Conservation Service has an index of published soil surveys listing by state all the county surveys that are available with date of publication. The current series of surveys are published with soil boundaries printed on a photomosaic base at a scale of 1:20,000 or 1:15,840, and are obtainable from the Government Printing Office, Washington, D. C.

State Extension Divisions. Practically all states have a bulletin on the soils of the state. Generally obtainable through the extension division of the state university or college.

THORP, JAMES. "Report on a Field Study of Soils of Australia, June 1954 to January 1955," *Science Bulletin* no. 1. Richmond, Indiana: Earlham College, 1957.

U.S. Bureau of Plant Industry, Soils, and Agricultural Engineering, *Soil Survey Manual, Agricultural Handbook* No. 18, 1951.

U.S. Department of Agriculture, *Yearbook of Agriculture,* 1938, "Soils and Men," Washington: 1938.

U.S. Department of Agriculture, *Yearbook of Agriculture,* 1957, "Soil," Washington: 1957.

U.S. Soil Conservation Service, Soil Survey Staff. *Soil Classification: A Comprehensive System, 7th Approximation.* Washington: 1960.

————. *Supplement to Soil Classification System (7th Approximation.)* Washington: 1967.

The 7th Approximation
of Soil Classification

An entirely new approach to soil classification was begun in 1951. The ideas of how this was to be accomplished went through a number of stages or "approximations," the 7th of which is now used in classifying the soils of the United States.

Soil Classification, A Comprehensive System, 7th Approximation was published in 1960 with a lengthy supplement added in 1967 (see bibliography). It has six levels of generalization which in descending sequence are: Order, Suborder, Great Group, Subgroup, Family, and Series. It is comprehensive as it is systematic. It considers the soil profile itself using generic factors of soil genesis, not climate nor vegetation. Figure 5.1 is a generalized map of the soil Orders and Suborders of the United States. A larger scale map in color is in the National Atlas and is available as a single sheet. Figure 5.2 is an illustration of the soils of the world as to the probable occurence of orders. This of course is tentative since very little field work has been done.

The order category is based on gross morphological characteristics of the soil profile such as color and organic matter content within general climatic parameters. The suborder permits fuller description of physical properties of the soils and such climatic variable as seasonality of precipitation or dryness that may affect certain physical properties. The great groups are more fully described as to specific profile characteristics such as the existence of a lime zone, an indurated layer, mottling, etc., as well as the arrangement of the horizons. Each great group, and there are about 120 of them, has a similar morphology.

Names of the ten orders end in *sol,* and each has a formative element abstracted from its name which is used as an ending for the names of the suborders, great groups, and subgroups. Alfisol, for example, is the name of one of the orders. Its formative element is *alf*. Names of suborders have two syllables, the first of which is suggestive of a property of the class and the final syllable is the formative element taken from the name of the order, as in aqualf. The two-syllable length is unique to the suborders. Great groups have names of more than two syllables formed by adding a prefix or more to the suborder name, as in Albaqualf. The suborder name makes up the final two syllables of every great group. Table 5.1 lists the names of the orders, the formative element, the derivation of the formative element and the mnemonicon and pronunciation of the formative element, and a brief definition of each order as well as the approximate great soil group equivalents of the old classification.

Table 5.2 lists the formative element in the names of suborders together with the derivation of the formative element, the mnemonicon, and the connotation of the formative element for each suborder.

Table 5.3 lists the formative elements for names of the great groups, the derivation of the formative element, the mnemonicon, and connotation of the formative element for each.

Take for example the great group *Fragiocrepts*. It is in the order *Inceptisols*, formative element *ept*; suborder *Ochrepts*, formative element *ochr* (G. base for *ochros*, pale; soils with little organic matter and light-colored surface, and never saturated with water). The prefix to the suborder name, *frag* (modified from L. *fragilis*, brittle, a brittle pan), completes the great group term. The complete definition of the great group would then incorporate the definitions of order, suborder, and the prefix used. Therefore, the Fragiocrepts are soils that have weakly differentiated horizons which show alteration of parent materials (Inceptisol definition); the parent materials are crystalline clay minerals; the solum contains a colored surface horizon and a B horizon that is low in minerals (definition of Ochrepts) in which a dense, non-brittle indurated horizon or fragipan has developed. Some of the Brown Podzolic and Gray-Brown Podzolic soils with fragipans would fit into this great group.

A subgroup of any great group may exist. It may have characteristics essentially those of one great group, yet have some properties of another class. If this is the case it will carry the name of the other in adjective form. Should the Fragiochrept typify the central concept of the group it is a Typic Fragiochrept. However, if it shows evidence of being poorly drained (a characteristic associated with wetness atypical of Ochrepts which are never saturated) yet has all

PATTERNS OF SOIL ORDERS AND SUBORDERS OF THE UNITED STATES

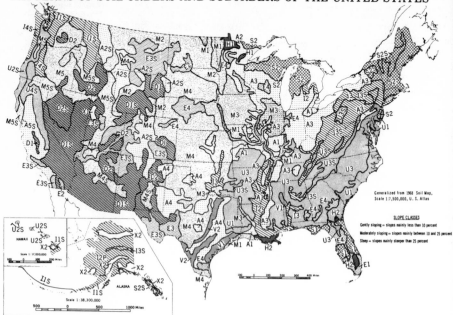

Generalized from 1968 Soil Map,
Scale 1:7,500,000, U. S. Atlas

SLOPE CLASSES

Gently sloping = slopes mainly less than 10 percent
Moderately sloping = slopes mainly between 10 and 25 percent
Steep = slopes mainly steeper than 25 percent

LEGEND

Only the dominant orders and suborders are shown. Each delineation has many inclusions of other kinds of soil. General definitions for the orders and suborders follow. For complete definitions see Soil Survey Staff, Soil Classification, A Comprehensive System, 7th Approximation, Soil Conservation Service, U. S. Department of Agriculture, 1960 (for sale by U. S. Government Printing Office) and the March 1967 supplement (available from Soil Conservation Service, U. S. Department of Agriculture). Approximate equivalents in the modified 1938 soil classification system are indicated for each suborder.

 ALFISOLS . . . Soils with gray to brown surface horizons, medium to high base supply, and subsurface horizons of clay accumulation; usually moist but may be dry during warm season

A1   AQUALFS (seasonally saturated with water) gently sloping; general crops if drained, pasture and woodland if undrained (Some Low-Humic Gley soils and Planosols)

A2   BORALFS (cool or cold) gently sloping; mostly woodland, pasture, and some small grain (Gray Wooded soils)

A2S   BORALFS steep; mostly woodland

A3   UDALFS (temperate or warm, and moist) gently or moderately sloping; mostly farmed, corn, soybeans, small grain, and pasture (Gray-Brown Podzolic soils)

A4   USTALFS (warm and intermittently dry for long periods) gently or moderately sloping; range, small grain, and irrigated crops (Some Reddish Chestnut and Red-Yellow Podzolic soils)

A5S   XERALFS (warm and continuously dry in summer for long periods, moist in winter) gently sloping to steep; mostly range, small grain, and irrigated crops (Noncalcic Brown soils)

 ARIDISOLS . . . Soils with pedogenic horizons, low in organic matter, and dry more than 6 months of the year in all horizons

D1   ARGIDS (with horizon of clay accumulation) gently or moderately sloping; mostly range, some irrigated crops (Some Desert, Reddish Desert, Reddish Brown, and Brown soils and associated Solonetz soils)

D1S   ARGIDS gently sloping to steep

D2   ORTHIDS (without horizon of clay accumulation) gently or moderately sloping; mostly range and some irrigated crops (Some Desert, Reddish Desert, Sierozem, and Brown soil and some Calcisols and Solonchak soils)

D2S   ORTHIDS gently sloping to steep

 ENTISOLS . . . Soils without pedogenic horizons

E1   AQUENTS (seasonally saturated with water) gently sloping some grazing

E2   ORTHENTS (loamy or clayey textures) deep to hard rock; gently to moderately sloping; range or irrigated farming (Regosols)

E3   ORTHENTS shallow to hard rock; gently to moderately slop mostly range (Lithosols)

E3S   ORTHENTS shallow to rock; steep; mostly range

E4   PSAMMENTS (sand or loamy sand textures) gently to moderately sloping; mostly range in dry climates, woodland or cropland in humid climates (Regosols)

 HISTOSOLS . . . Organic soils

H1   FIBRISTS (fibrous or woody peats, largely undecomposed) mostly wooded or idle (Peats)

H2   SAPRISTS (decomposed mucks) truck crops if drained, idle undrained (Mucks)

 INCEPTISOLS . . . Soils that are usually moist, with pedogenic horizons of alteration of parent materials but not of accumulation

I1S   ANDEPTS (with amorphous clay or vitric volcanic ash and pumice) gently sloping to steep; mostly woodland; in Hawaii mostly sugar cane, pineapple, and range (Ando soils, some Tundra soils)

QUEPTS (seasonally saturated with water) gently sloping; if drained, mostly row crops, corn, soybeans, and cotton; if undrained, mostly woodland or pasture (Some Low-Humic Gley soils and Alluvial soils)

QUEPTS (with continuous or sporadic permafrost) gently sloping to steep; woodland or idle (Tundra soils)

CHREPTS (with thin or light-colored surface horizons and little organic matter) gently to moderately sloping; mostly pasture, small grain, and hay (Sols Bruns Acides and some Alluvial soils)

CHREPTS gently sloping to steep; woodland, pasture, small grains

UMBREPTS (with thick dark-colored surface horizons rich in organic matter) moderately sloping to steep; mostly woodland (Some Regosols)

MOLLISOLS . . . Soils with nearly black, organic-rich surface horizons and high base supply

AQUOLLS (seasonally saturated with water) gently sloping; mostly drained and farmed (Humic Gley soils)

BOROLLS (cool or cold) gently or moderately sloping, some steep slopes in Utah; mostly small grain in North Central States, range and woodland in Western States (Some Chernozems)

UDOLLS (temperate or warm, and moist) gently or moderately sloping; mostly corn, soybeans, and small grains (Some Brunizems)

USTOLLS (intermittently dry for long periods during summer) gently to moderately sloping; mostly wheat and range in western part, wheat and corn or sorghum in eastern part, some irrigated crops (Chestnut soils and some Chernozems and Brown soils)

USTOLLS moderately sloping to steep; mostly range or woodland

XEROLLS (continuously dry in summer for long periods, moist in winter) gently to moderately sloping; mostly wheat, range, and irrigated crops (Some Brunizems, Chestnut, and Brown soils)

XEROLLS moderately sloping to steep; mostly range

SPODOSOLS . . . Soils with accumulations of amorphous materials in subsurface horizons

AQUODS (seasonally saturated with water) gently sloping; mostly range or woodland; where drained in Florida, citrus and special crops (Ground-Water Podzols)

ORTHODS (with subsurface accumulations of iron, aluminum, and organic matter) gently to moderately sloping; woodland, pasture, small grains, special crops (Podzols, Brown Podzolic soils)

ORTHODS steep; mostly woodland

ULTISOLS . . . Soils that are usually moist, with horizon of clay accumulation and a low base supply

AQUULTS (seasonally saturated with water) gently sloping; woodland and pasture if undrained, feed and truck crops if drained (Some Low-Humic Gley soils)

HUMULTS (with high or very high organic-matter content) moderately sloping to steep; woodland and pasture if steep, sugar cane and pineapple in Hawaii, truck and seed crops in Western States (Some Reddish-Brown Lateritic soils)

U3 UDULTS (with low organic-matter content; temperate or warm, and moist) gently to moderately sloping; woodland, pasture, feed crops, tobacco, and cotton (Red-Yellow Podzolic soils, some Reddish-Brown Lateritic soils)

U3S UDULTS moderately sloping to steep; woodland, pasture

U4S XERULTS (with low to moderate organic-matter content, continuously dry for long periods in summer) range and woodland (Some Reddish-Brown Lateritic soils)

VERTISOLS . . . Soils with high content of swelling clays and wide deep cracks at some season

V1 UDERTS (cracks open for only short periods, less than 3 months in a year) gently sloping; cotton, corn, pasture, and some rice (Some Grumusols)

V2 USTERTS (cracks open and close twice a year and remain open more than 3 months); general crops, range, and some irrigated crops (Some Grumusols)

AREAS with little soil . . .

X1 Salt flats

X2 Rock land (plus ice fields in Alaska)

NOMENCLATURE

The nomenclature is systematic. Names of soil orders end in *sol* (L. *solum*, soil), e. g., ALFISOL, and contain a formative element used as the final syllable in names of taxa in suborders, great groups, and subgroups.

Names of suborders consist of two syllables, e. g., AQUALF. Formative elements in the legend and their connotations are as follows:

and	— Modified from Ando soils; soils from vitreous parent materials
aqu	— L. *aqua*, water; soils that are wet for long periods
arg	— Modified from L. *argilla*, clay; soils with a horizon of clay accumulation
bor	— Gr. *boreas*, northern; cool
fibr	— L. *fibra*, fiber; least decomposed
hum	— L. *humus*, earth; presence of organic matter
ochr	— Gr. base of ochros, pale; soils with little organic matter
orth	— Gr. *orthos*, true; the common or typical
psamm	— Gr. *psammos*, sand; sandy soils
sapr	— Gr. *sapros*, rotten; most decomposed
ud	— L. *udus*, humid; of humid climates
umbr	— L. *umbra*, shade; dark colors reflecting much organic matter
ust	— L. *ustus*, burnt; of dry climates with summer rains
xer	— Gr. *xeros*, dry; of dry climates with winter rains

Figure 5.1. Patterns of Soil Orders and Suborders of the United States according to the Seventh Approximation.

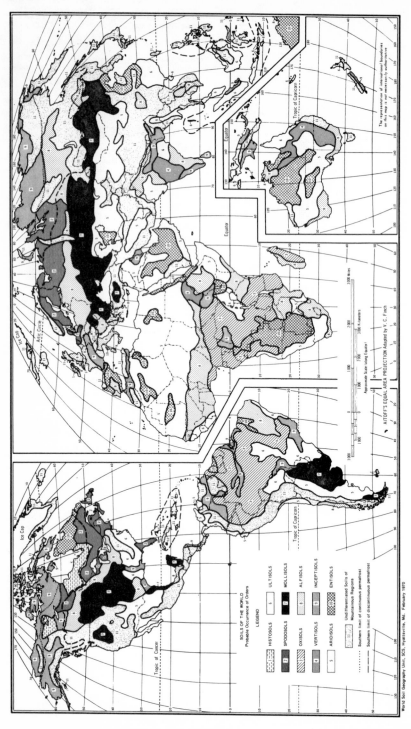

Figure 5.2. Soils of the World—Probable Occurrence of Orders.

SOILS OF THE WORLD
Probable Occurrence of Orders

LEGEND

	6	ULTISOLS
HISTOSOLS		
	7	MOLLISOLS
SPODOSOLS		
	8	ALFISOLS
OXISOLS		
	9	INCEPTISOLS
VERTISOLS		
4		
ARIDISOLS		ENTISOLS
5		

11 — Undifferentiated Soils of Mountainous Regions

Southern limit of continuous permafrost

•••••••• Southern limit of discontinuous permafrost

AITOFF'S EQUAL AREA PROJECTION Adapted by V. C. Finch

The representation of international boundaries on this map is not necessarily authoritative.

World Soil Geography Unit, SCS, Hyattsville, Md., February 1970

76

The names of the Orders, the formative element, the derivation of the formative element and the mnemonic and pronunciation of the formative element, and a brief definition of each order as well as the approximate Great Soil Group equivalents of the old classification.

Order Name	Definition	Derivation of formative element	Pronunciation of formative element*	Approximate equivalents in old system
Entisols	soils without pedogenic horizons	Nonsense symbol	*recent*	Azonal, some Low-Humic Gley soils
Vertisols	cracking clay soils	L. *verto*, turn	*invert*	Grumusols
Inceptisols	soils with weakly differentiated horizons showing alteration of parent materials	L. *inceptum*, beginning	*inception*	Ando, Sol Brun Acid, some Brown Forest, Low-Humic Gley, and Humic Gley soils
Aridisols	soils with pedogenic horizons, low in organic matter, usually dry	L. *aridus*, dry	*arid*	Desert, Reddish Desert, Sierozem, Solonchak, some Brown and Reddish Brown soils and associated Solonetz
Mollisols	soils with nearly black, organic-rich surface horizon and high base supply	L. *mollis*, soft	*mollify*	Chestnut, Chernozem, Brunizem (Prairie), Rendzinas, some Brown, Brown Forest, and associated Solonetz, and Humic Gley soils
Spodosols	soils have that accumulation of amorphous materials in subsurface horizons	Gk. *spodos*, wood ash	*Podzol*; odd	Podzols, Brown Podzolic soils, and Ground-water Podzols
Alfisols	soils with gray to brown surface horizon, medium to high base supply, and subsurface horizons of clay accumulation	Nonsense symbol	*Pedalfer*	Gray-Brown Podzolic soils, Gray Wooded soils, Non-Calcic Brown soils, Degraded Chernozems, and associated Planosols and some Half-Bog soils
Ultisols	soils with horizons of clay accumulation and low base supply	L. *ultimus*, last	*ultimate*	Red-Yellow Podzolic soils, Reddish-Brown Lateritic soils of the U.S., and associated Planosols and Half-Bog soils
Oxisols	soils that have mixtures principally of kolin, hydrated oxides, and quartz	F. *oxide*, oxide	*oxide*	Laterite soils, Latosols
Histosols	organic (peat and muck) soils	GK. *histos*, tissue	*histology*	Bog soils

TABLE 5.2

The formative element in the names of Suborders together with the derivation of the formative element, the mnemonicon, and the connotation of the formative element for each suborder.

Formative Elements in Names of Suborders

Formative elements	Derivation of formative element	Mnemonicon	Connotation of formative element
alb	L. *albus,* white	albino	Presence of albic horizon (a bleached eluvial horizon)
and	Modified from *Ando*	Ando	Ando-like
aqu	L. *aqua,* water	aquarium	Characteristics associated with wetness
ar	L. *arare,* to plow	arable	Mixed horizons
arg	Modified from argillic horizon; L. *argilla,* white clay	argillite	Presence of argillic horizon (a horizon with illuvial clay)
bor	Gr. *boreas,* northern	boreal	Cool
ferr	L. *ferrum,* iron	ferruginous	Presence of iron
fibr	L. *fibra,* fiber	fibrous	Least decomposed stage
fluv	L. *fluvius,* river	fluvial	Flood plains
hem	Gr. *hemi,* half	hemisphere	Intermediate stage of decomposition
hum	L. *humus,* earth	humus	Presence of organic matter
lept.	Gr. *leptos,* thin	leptometer	Thin horizon
ochr	Gr. base of *ochros,* pale	ocher	Presence of ochric epipedon (a light-colored surface)
orth	Gr. *orthos,* true	orthophonic	The common ones
plag	Modified from Ger. *plaggen,* sod		Presence of plaggen epipedon
psamm	Gr. *psammos,* sand	psammite	Sand textures
rend	Modified from Rendzina	Rendzina	Rendzina-like
sapr	Gr. *sapros,* rotten	saprophyte	Most decomposed stage
torr	L. *torridus,* hot and dry	torrid	Usually dry
trop	Modified from Gr. *tropikos,* of the solstice	tropical	Continually warm
ud	L. *udus,* humid	udometer	Of humid climates
umbr	L. *umbra,* shade	umbrella	Presence of umbric epipedon (a dark-colored surface)
ust	L. *ustus,* burnt	combustion	Of dry climates, usually hot in summer
xer	Gr. *xeros,* dry	xerophyte	Annual dry season

Soil Surface Staff, Soil Conservation Service *Supplement to Soil Classification System* (The Approximation) 1967.

TABLE 5.3

The formative elements for names of the Great Groups, the derivation of the formative element, the mnemonicon, and connotation of the formative element for each.

Formative Elements for Names of Great Groups

Formative element	Derivation of formative element	Mnemonicon	Connotation of formative element
acr	Modified from Gr. *akros*, at the end.	acrolith	Extreme weathering
agr	L. *ager*, field	agriculture	An agric horizon
alb	L. *albus*, white	albino	An albic horizon
and	Modified from *Ando*	Ando	Ando-like
anthr	Gr. *anthropos*, man	anthropology	An anthropic epipedon
aqu	L. *aqua*, water	aquarium	Characteristic associated with wetness
arg	Modified from argillic horizon; L. *argilla*, white clay	argillite	An argillic horizon
calc	L. calcis, *lime*	calcium	A calcic horizon
camb	L. *cambiare*, to exchange	change	A cambic horizon
chrom	Gr. *chroma*, color	chroma	High chroma
cry	Gr. *kryos*, coldness	crystal	Cold
dur	L. *durus*, hard	durable	A duripan
dystr, dys	Modified from Gr. dys, ill; *dystrophic*, infertile	dystrophic	Low base saturation
eutr, eu	Modified from Gr. *eu*, good; *eutrophic* fertile	eutrophic	High base saturation
ferr	L. *ferrum*, iron	ferric	Presence of iron
frag	Modified from L. *fragilis*, brittle	fragile	Presence of fragipan
fragloss	Compound of *fra(g)* and *gloss*		See the formative elements *frag* and *gloss*
gibbs	Modified from *gibbsite*	gibbsite	Presence of gibbsite
gloss	Gr. *glossa*, tongue	glossary	Tongued
hal	Gr. *hals*, salt	halophyte	Salty
hapl	Gr. *haplous*, simple	haploid	Minimum horizon
hum	L. *humus*, earth		Presence of humus
hydr	Gr. *hydor*, water	hydrophobia	Presence of water
hyp	Gr. *hypnon*, moss	hypnum	Presence of hypnum moss
luo, lu	Gr. *louo*, to wash	ablution	Illuvial
moll	L. *mollis*, soft	mollify	Presence of mollic epipedon
nadur	Compound of *na(tr)*, and *dur*		
natr	Modified from *natrium*, sodium		Presence of natric horizon
ochr	Gr. base of *ochros*, pale	ocher	Presence of ochric epipedon (a light-colored surface)
pale	Gr. *paleos*, old	paleosol	Old development
pell	Gr. *pellos*, dusky		Low chroma
plac	Gr. base of *plax*, flat stone		Presence of a thin pan
plag	Modified from Ger. *plaggen*, sod		Presence of plaggen horizon
plinth	Gr. *plinthos*, brick		Presence of plinthite
quartz	Ger. *quarz*, quartz	quartz	High quartz content
rend	Modified from Rendzina	Rendzina	Rendzina-like
rhod	Gr. base of *rhodon*, rose	rhododendron	Dark-red colors
sal	L. base of *sal*, salt	saline	Presence of salic horizon
sider	Gr. *sideros*, iron	siderite	Presence of free iron oxides
sphagno	Gr. *sphagnos*, bog	sphagnum-moss	Presence of sphagnum-moss
torr	L. *torridus*, hot and dry	torrid	Usually dry
trop	Modified from Gr. *tropikos*, of the solstice	tropical	Continually warm
ud	L. *udus*, humid	udometer	Of humid climates
umbr	L. base of *umbra*, shade	umbrella	Presence of umbric epipedon
ust	L. base of *ustus*, burnt	combustion	Dry climate, usually hot in summer
verm	L. base of *vermes*, worm	vermiform	Wormy, or mixed by animals
vitr	L. *vitrum*, glass	vitreous	Presence of glass
xer	Gr. *xeros*, dry	xerophyte	Annual dry season
sombr	F. *sombre*, dark	somber	A dark horizon

the other characteristics of the Ochrepts it would be classified as an Aquic Fragiochrept (Table 5.4). Any of the formative elements from the above tables or the names of orders, suborders, or great groups may be used as a modifying adjective to identify a subgroup or "intergrade subgroup" of the great group. Thus intergrade subgroups toward other great groups in the *same suborder*, in a *different suborder*, or *other orders* are possible.

TABLE 5.4
Example of Order, Suborder, Great Group and Subgroup.

Inc*ept*isol	Order
Ochr*ept*	Suborder
Fragiochr*ept*	Great group
Aquic Fragiochrept	Subgroup

*Italicized letters are formative element in name of order.

Subgroups not intergrading toward any known kind of soil are called "extragrades" and show properties not characteristic of a class in any order. The formative elements together with the meaning of each is given in Table 5.5. In use a subgroup that has broken horizons, for example, would have its name preceded by *Ruptic*. Those that are truncated by hard rock or shallow to rock are *Lithic* subgroups and so on.

In a broad overall view many soils seem very similar in the surface horizons. They may be dark brown or black in color as are the Brunizems and the Chernozems. As a quick descriptive means of identifying broad characteristics both of surface and subsurface horizons, a diagnosis of readily recognizable characteristics, certain "diagnostic" horizons are recognized. Surface horizons, not necessarily the A horizon alone, are called "epipedons" (Gk. *epi*, over; *pedon*, soil). In somewhat abbreviated form they are as follows:

DIAGNOSTIC SURFACE HORIZONS

Mollic epipedon. A surface horizon containing a thick, dark surface layer saturated with bivalent cations is a mollic epipedon. It has a narrow carbon to nitrogen ratio and a moderate to strong structure.
Umbric epipedon. It is like the Mollic epipedon except that much of the dominant exchangeable cations are hydrogen as may be determined in the laboratory.

TABLE 5.5

The names of subgroups, extragrades, together with their derivations.

Formative element	Derivation of formative element	Mnemonicon	Connotation of formative element
abruptic	L. *abruptum,* torn off	abrupt	Abrupt textural change
allic	Modified from *aluminum*		Presence of extractable aluminum
arenic	L. *arena,* sand	arenose	Sandy texture
clastic	Gr. *klastos,* broken	clastic	High mineral content
cumulic	L. *cumulus,* heap	accumulation	Thickened epipedon
glossic	Gr. *glossa,* tongue	glossary	Tongued
gross-arenic	L. *grossus,* thick, and L. *arena,* sand		Thick sandy layer
limnic	Modified from Gr. *limn,* lake	limnology	Presence of a limnic layer
lithic	Gr. *lithos,* stone	lithosphere	Presence of a lithic contact
leptic	Gr. *leptos,* thin		A thin solum
pergelic	L. *per,* throughout in time and space, and L. *gelare,* to freeze		Permanently frozen or having permafrost
petro-calcic	Gr. *petra,* rock and calcic from calcium		Petrocalcic horizon
plinthic	Modified from Gr. *plinthos,* brick	plinthite	Presence of plinthite
ruptic	L. *ruptum,* broken	rupture	Intermittent or broken horizons
stratic	L. *stratum,* a covering	stratified	Stratified layers
superic	L. *superare,* to overtop	superimpose	Presence of plinthite in the surface
pachic	Gr. *pachys,* thick	pachyderm	A thick epipedon

Anthropic epipedon. It is similar to the Mollic epipedon but is higher in phosphates. The anthropic epipedon is formed under long continuous farm operations, involving the additions of large amounts of organic matter, nitrogen, and phosphates. In a sense it is a man-made soil since it has been worked again and again to plow depth and has had much organic matter and fertilizers added by man. Supposedly this is largely European and has been identified conforming to field lines.

Histic epipedon. A horizon characterized by a thin organic horizon of virgin peaty materials, or by a very high amount of organic material, a result of mixing of peat and mineral soil on plowing. It is usually wet or may be artificially drained.

Ochric epipedon. This horizon is light in color, low in organic matter or very thin overall.

Plaggen epipedon. The surface horizon has been modified by its long continued use as a meadow, and manuring has given it a high organic matter content. In it a wide variety of artifacts may be found.

DIAGNOSTIC SUBSURFACE HORIZONS

Argillic horizon. A horizon in which a high percentage of silicate clays have accumulated. In time, erosion may cause it to be exposed on the surface. Much of the clay is secondary clay and therefore formed in place from parent materials rather than being illuviated.

Agric horizon. This is an illuvial horizon formed under cultivation. Many changes are observable in the soil flora and fauna as a result of man's manipulations of the soil.

Natric horizon. This is like the above except that its structure is prismatic or columnar. It is high in exchangeable sodium and is common to Solonetz soils.

Sodic horizon. An illuvial layer of free sesquioxides and organic carbon, free iron and illuvial crystalline clay. Some may be cemented to form an ortstein layer.

Cambic horizon. An altered horizon in which little evidence remains of the original rock structure. They are common in transported materials.

Oxic horizon. A subsurface horizon which contains the hydrated oxides of iron and aluminum. They are seldom found except in the tropical and subtropical climates at relatively low elevations.

Calcic horizon. Those horizons that contain accumulations of calcium and commonly developed in unconsolidated materials.

The Soils in the United States According to the 7th Approximation

The general distinction is made between warm and cool soils with a mean annual soil temperature of approximately 47° F (8° C). The warm soils have a mean annual soil temperature that is greater than 47° and within this temperature parameter soils are considered moist, wet or dry. The cool soils are those whose mean annual temperature is less than 47° F and can be moist, or wet. These general climatic characteristics are brought out in the general descriptions of the major categories, order, suborder, and great group that follow. Also, in very general terms the dominant land use of the soils is given with each suborder. For identification with the older system of soil classification the name or names of the Great Soil Groups as used prior to 1960 is given in parenthesis following the name of each Great Group of the 7th Approximation.

While still of a tentative nature, soil temperature classes for differentiation of families in all orders are given in Table 5.6. The soil temperature range between summer and winter and the mean annual soil temperature are key factors in the definitions.

Other physical properties are taken into consideration as well as temperature for soil family names. As yet they are ill defined and will not be discussed further.

TABLE 5.6
Soil Temperature Classes.

Soil Temperature Classes
(50 cm depth)

Mean Annual Soil Temperature	
with annual soil temper- range less than 5° C	with annual soil temp. range* 5° C or greater
Isofrigid ------------------------------------- <8°C (47°F) --------------------- Frigid	
Isomesic ------------------------------- 8° to 15° (47-59°F)------------------ Mesic	
Isothermic ---------------------------- 15° to 22° (59-72°F) ---------------- Thermic	
Isohyperthermic --------------------- >22°C (72°F) --------------------- Hyperthermic	

*between mean summer (J J A) and mean winter (D J F)

The soil descriptions that follow are brief and point out major characteristics only. In most cases an understanding of the meanings of formative elements is sufficient to identify each.

alfisols
The alfisols are medium to high in bases with a base saturation at pH 8.2. The profile exhibits a gray to brown surface horizon and subsurface horizons of clay accumulation. Although found under a wide range of climates (some with a mean annual soil temperature above 47° and some below) they are nevertheless usually moist. During the warm season of the year droughtiness is characteristic of some.

Aqualfs
The formative element, *aqua* (Latin for water) is indicative of wetness throughout the profile. Aqualfs are seasonally wet Alfisols containing mottles, iron-manganese concretions, or gray colors. Where the soils are drained they are used for general crops and where undrained, for pasture and woodland. The major areas of Aqualfs are central Missouri, southern Illinois, northwestern Ohio, and southeast Michigan.

Albaqualfs (Planosols). This great group of Aqualfs has a bleached white upper horizon that changes abruptly to the heavier textured B horizon of clay accumulation.

Fragiaqualfs (Low-Humic Gley soils with a fragipan). A great group of Aqualfs that have a dense brittle but not indurated horizon or fragipan in the subsoil.

Glossaqualfs (Planosols). These soils are Aqualfs that have "tongues" of an upper bleached, white horizon reaching down and into the underlying B horizon of clay accumulation.

Natraqualfs (Solonetz soils). Aqualfs that have a subsurface horizon of clay accumulation in which the sodium (alkali) content is high, consequently a soil of the deserts.

Ochraqualfs (Low-Humic Gley soils). Aqualfs with a light colored surface horizon that changes gradually in texture downward into the B horizon.

Boralfs

The prefix to this group of Alfisols, *bor* (from the Greek *boreas*), means northern or cool. Boralfs are those Alfisols of cool to cold regions of the middle latitudes and higher altitudes. Mean summer soil temperatures may range from less than 47° F to 59° F. Their predominant use is for woodland, pasture, and some of the small grains. In the United States they are principally soils of the Rocky Mountain areas of Colorado, Wyoming and Montana.

Cryoboralfs (Gray Wooded soils). These Boralfs have developed under cold climatic conditions of intermediate to near high latitudes.

Eutroboralfs (Gray Wooded soils). These have a B horizon in which the accumulation of bases is high and that is climatically dry for short periods in most years. Such conditions are found on the inner margin of the Humid Continental climate.

Fragiboralfs (Gray Wooded soils with a fragipan). These Boralfs have a B horizon of clay accumulation within two feet of the surface and in which there is found a dense brittle but not indurated horizon or fragipan.

Glossoboralfs (Gray Wooded soils). These are Boralfs in which extensions of the upper white, bleached A horizon are found in the B horizon of clay accumulation.

Udalfs

The formative element, *ud* (Latin, *udus*), humid, is indicative of their location in the humid climates. They are found in the Humid Subtropical as well as the cooler Humid Continental climates. The soils are usually moist during the warm season of the year but may be dry in some horizons for brief periods during the warm season. They find their greatest use for row crops, small grains, and pastureland. For the most part these are found on glacial materials in Minnesota, Wisconsin, Indiana, Michigan and Ohio; and on sedimentary materials in Tennessee and Mississippi.

Fragiudalfs (Gray-Brown Podzolic soils with a fragipan). Soils of the Humid Continental climate areas, they have a dense brittle, but not indurated, horizon of fragipan found usually below the horizon of clay accumulation.

Hapludalfs (Gray-Brown Podzolic soils without the fragipan formation). This group of Udalfs has minimal horizon development in which a subsurface horizon of clay accumulation is relatively thin or is brownish in nature.

Paleudalfs (Red-Yellow Podzolic and Gray-Brown Podzolic soils). These are old soils that have a thick B horizon of clay accumulation.

Ustalfs

The prefix *ust*, (*ustus*, Latin for burnt), connotes high temperature conditions. The Ustalfs are Alfisols that are found in the temperate to subtropical and tropical climatic areas under rather dry conditions. Summer rains are the rule. The soils are reddish brown in color. During dry periods of some duration agricultural use is limited. They are used for range, small grains, and irrigated crops. The inner coastal plain of Texas and the Great Plains of western Texas are the major areas of Ustalfs.

Haplustalfs (Reddish Chestnut and Reddish Brown soils). Formed under subhumid to semiarid climatic conditions, the Haplustalfs have a relatively thin, brownish, B horizon of clay accumulation.

Paleustalfs (Reddish Chestnut and Reddish Brown soils). These Ustalfs have developed an indurated (petrocalcic) horizon in which the cementing materials are carbonates. The B horizon is either a thick reddish clay accumulation or clayey in the upper B grading abruptly to a lighter texture in the A horizon above.

Rhodustalfs (Reddish Prairie soils). They have developed under slightly more moist conditions than the Paleustalfs or the Haplustalfs. The B horizon is reddish; however, unlike the Paleustalfs or the Haplustalfs, the clay accumulation is relatively thin.

Xeralfs

Xer (Greek for *xeros*,) dry, is indicative of a dry climatic environment. The Xeralfs are found in the Dry-Summer Subtropical climatic areas where the rain that is received is a wintertime phenomenon. During the summers these soils are continually dry for long periods, necessitating irrigation for some crops. Other than for irrigated crops, these soils are used for range and small grains. Coastal, southern California and much of the Great Valley of California contain Xeralfs.

Durixeralfs (Noncalcic Brown soils containing a hardpan). These Xeralfs have a hardpan or durapan that is cemented with silica. Because of the presence of the hardpan it is of limited use agriculturally. *Haploxeralfs* (Noncalcic Brown soils). The Haploxeralfs are Noncalcic Brown soils of rather simple development. Beneath the A horizon is a relatively thin B horizon of clay accumulation that may be brownish in color.

Palexeralfs (Noncalcic Brown soils). The B horizon contains an indurated or petrocalcic horizon that has carbonates as the cementing material and may have either a thick reddish layer of clay accumulation or clay that is more widely spread in the upper B horizon, above which the A horizon is of a lighter texture.

aridisols

The Aridisols develop in the Desert and Steppe climates. They are low in organic matter and the profile is dry for seven months or more.

Argids

The prefix *arg* is a modification of the Latin *argilla*, meaning "clay." The Argids are those Aridisols that have a B horizon of clay accumulation in which sodium may or may not be present. They are used extensively as range land and for irrigated crops where water is available. The major areas in the United States which contain Argids are the Great Basin of Nevada, California, Arizona, New Mexico and Texas, the Wyoming Basin, and the Great Plains of northeast Wyoming and southeast Colorado.

Durargids. (Desert, Red Desert, Sierozem, and some Brown soils containing a hardpan). Silicious materials form the cement for the hardpan (duripan) found in these desert soils.

Haplargids (Sierozem, Desert, Red Desert, and some Brown soils). These are Argids without the hardpan found in the Duragids. They have a loamy B horizon of clay accumulation and may or may not have an accumulation of alkali (sodium).

Nadurargids (Solonetz soils containing a hardpan). The B horizon of the Nadurargids contains sodium and overlies a hardpan formed by silica cementation.

Natrargids (Solonetz soils). These are similar to the Nadurargids in that they have a horizon of clay and alkali accumulation and dissimilar in that no hardpan is formed.

Paleargids (Sierozems, Desert and Red Desert soils). The B horizon is of a clay texture that may or may not contain sodium (alkali). Paleargids may have an indurated horizon cemented by carbonates (petro-calcic horizon) in some cases.

Orthids

The prefix *orth* comes from the Greek *orthos*, meaning "true." The Orthid subgroup reflects the environmental characteristic of calcification. The Orthids are Aridisols that have accumulations of calcium carbonate, gypsum, or other salts more soluble than gypsum, but do not have a horizon of clay accumulation. They may have horizons from which some materials have been removed or altered and are utilized as rangeland and for some irrigated crops when water is available. The Orthids are found in northwestern Nevada, the Snake River valley of Idaho, portions of the Colorado Plateau of Arizona and Utah, central New Mexico and western Texas.

Calciorthids (Calcisols). These contain large amounts of calcium carbonate or gypsum in the B horizon. Some former lithosols in which the depth is less than 20 inches to bedrock are contained in the Calci-orthid group of soils.

Camborthids (Sierozems, Desert, and Red Desert soils). The Cam-borthids have had materials removed or altered within the profile. There is no accumulation of calcium carbonate or gypsum in large amounts.

Durorthids (Regosols, Calcisols, or Alluvial soils containing a hard-pan). The hardpan or durapan of the Durorthids is one cemented with silica, hence quite hard.

Paleorthids (Calcisols). Paleorthids contain a hardpan which in this case is cemented with carbonates (petrocalcic horizon) and not likely to be as hard and impenetrable as the durapan of the above.

Salorthids (Solonchaks). These are Orthids in which large amounts of salts other than sodium have accumulated in the B horizon. These salts are more easily removed where irrigation water is adequate.

entisols

The Entisols are found in a wide range of climatic types and have no pedogenic horizon development.

Aquents

The prefix *aqua*, from the Latin for "water," is indicative of continued wetness of these soils for periods of varying length of time. The Aquents are soils that are seasonally or permanently wet. The profile is grayish and contains mottles. They are used to a limited extent for pasture and are largely confined to southern coastal Florida.

Haplaquents (Low-Humic Gley soils). Haplaquents are soils that have a rather limited profile development, a result of the wet environment in which they are found. They have a loamy fine sand or finer texture.

Hydraquents. These are Aquents that are permanently wet and have very fine textural characteristics. They have little load bearing ability to the point that they will not support the weight of livestock, consequently are of little use agriculturally.

Psammaquents (Low-Humic Gley soils and some poorly drained Regosols). The Psammaquents are those Aquents of sandy textures. They are loamy fine sands or coarser textures unlike the Haplaquents which range downward from the loamy fine sands to finer textures.

Fluvents

The formative element *fluv* is from the Latin for "river," hence are floodplain soils. The Fluvents are formed of alluvium and were classified as alluvial soils. Generally of a loamy or clayey texture, the Fluvents contain organic matter in variable amounts with the greatest amount close to the surface.

Torrifluvents are alluvial soils that are dry for periods of nine months or more.

Udifluvents. The prefix *ud* (Latin *udus,* humid) indicates that these are alluvial soils of humid climates and moist at all times, very unlike the Torrifluvents.

Xerofluvents are those Fluvents that are in summer-dry climates (Mediterranean), the profiles of which are seasonally dry for relatively long periods of time.

Orthents

The formative element *orth* is from the Greek and is defined as "true." The Orthents are then typical soils of the Entisols order, having no horizon development. These are loamy or clayey Entisols that have a regular decrease in organic matter content with depth. They are used for range or irrigated crops in dry regions and for general farming in humid regions. Orthents are most extensive on the Montana plains but are also found in southwest Texas, the Colorado Plateau of Colorado, Utah, Arizona, and New Mexico as well as a few other scattered areas.

Cryorthents (Alluvial soils, Regosols, and Lithosols). These Orthents are soils of the colder latitudes and higher altitudes of the United States.

Torriorthents (Regosols). These are Orthents that are dry for periods of time in excess of 9 months. Some Torriorthents may be shallower than 20 inches to bedrock. The latter are former Lithosols.

Udorthents (Regosols). Orthents that are found in the humid climatic areas and are usually moist, comprise this group.

Ustorthents (Regosols). These are Orthents found in hot, dry summer climates and are intermittently dry for long periods of time during the warm season. Some have depths that are less than 20 inches to bedrock. These were former Lithosols.

Xerorthents (Regosols, Brown, or Alluvial soils). The Xerorthents are soils of the Dry-Summer Subtropical climate. During the dry summer they are constantly dry for long periods of time. Crops on these soils require irrigation. Former Lithosols where the soil is less than 20 inches to bedrock are included in this great group.

Psamments

The connotation of the formative element, *psamm,* is sandy texture. The Psamments then are Entisols that have textures of loamy fine sand or coarser. They find their greatest usage in rangeland, wild hay production and some hardy vegetables in Alaska, woodland and small grains where warm and moist, pasture and citrus in Florida, and range and irrigated crops where the climate is warm and dry. The largest single area is in the Sand Hills area of Nebraska. Small areas are found in Colorado, Texas, Florida, Wisconsin, and Montana.

Cryopsamments (Regosols) are Psamments of the cold climates.

Quartzipsamments (Regosols) are Psamments that consist almost entirely of the mineral quartz.

Torripsamments (Regosols) are Psamments of dry climates. They have developed on parent materials of a nonresistant character and are easily weathered. The climate in which they develop is so dry that available moisture is insufficient to keep the soil moist for as long as three consecutive months.

Udipsamments (Regosols) are Psamments that, like the Torripsamments, contain easily weatherable minerals. They differ, however, in that the profile is usually moist in most years.

Ustipsamments also contain easily weatherable minerals. Because of summer rainfall and winter dryness the soils are intermittently dry for long periods. Some of the Regosols belong in this group.

Xeropsamments are Psamments that have developed in climates with rainy winters and dry summers. The Dry-Summer Subtropical (Mediterranean) climate is typical of the environment in which Xeropsamments are found.

histosols

The Histosols are the peat and muck soils found over a wide range of latitude. They include soils in which the decomposition of plant residues ranges from slight to great. All are formed in marshes, are used for woodland, or simply lie idle. Where the land has been drained truck crops will be found. Histosols of cool regions (peat) are largely made up of undecomposed plant residues. Histosols of warm regions (peat or muck) consist largely of moderately to highly decomposed plant residues. Histosols are found in the Everglades of Florida, the Louisiana delta and marsh areas of northern Minnesota. Many minor areas are widely scattered.

inceptisols

Inceptisols are soils that have weakly differentiated horizons. Weathering has altered or removed soil materials but there has been little accumulation of these materials in the profile. These soils are usually moist, but during the warm season of the year some are dry for part of the time.

Andepts

The formative element *and* is modified from one of the Great Soil Groups of the older classification, the Ando soils, which are those formed from volcanic materials. The Andepts, then, are those Inceptisols that have formed in pyroclastic parent material or have a low bulk density and large amounts of amorphous materials, or both. They are soils used largely for woodland and range or pasture; and are found for the most part in the northern Rocky Mountains of Idaho and the Cascade Mountains of Washington.

Cryandepts (Brown Podzolic or Gray-Brown Podzolic soils) are the Inceptisols of cold climatic areas, largely high alititudinal locations.

Dystrandepts (the Ando soils) are Andepts that have a thick dark-colored surface horizon (light-colored in some areas) that have only small amounts of bases within the profile largely a result of the parent materials from which they developed.

Eutrandepts (Brown or Reddish Brown soils). The prefix *eutr* is modified from eutrophic, fertile. Thus the Eutrandepts are Andepts that have a high base saturation. They are dark-colored and have a thick surface horizon because of their more favorable topographic position.

Hydrandepts (Hydrol Humic Latosols) are Andepts that harden permanently if they become dried. They are reported only in Hawaii in perhumid (excessively wet) climates. *Vitrandepts* (Regosols) are Andepts formed in pumice or slightly weathered volcanic ash.

Aquepts

The formative element *aqu* is indicative of the wetness that exists in these Inceptisols. They are seasonally saturated, have an organic surface horizon, and are mottled or may be of a gray color. Some in the high latitudes may have permafrost. They are used for pasture, hay, and in Alaska when drained have a limited use for hardy vegetables. In the Southeastern United States they are used for woodland pasture or, if drained, for row crops. The range of climates in which the Aquepts are found is wide. In addition to covering much of Alaska the Aquepts are found extensively on the alluvial lands of the Mississippi Valley.

Andaquepts (Humic-Gley and Alluvial soils). These are Aquepts that have formed in ashy (vitric pyroclastic) materials, or have low bulk density and large amounts of amorphous materials, or both.

Cryaquepts (Tundra soils) are Aquepts of the cold Arctic climatic regions.

Fragiaquepts (Low-Humic Gley soils with fragipan). These Aquepts have a dense brittle but not indurated horizon or fragipan.

Haplaquepts (Low-Humic Gley soils) are Aquepts that have little horizon development. They are thin and may range in color from gray to black.

Humaquepts (Humic-Gley soils). These have sufficient humus present to darken the color of the surface horizon. Humaquepts are acid in reaction.

Ochrepts

The prefix *ochr* comes from the Greek *ochros,* meaning "pale." The Ochrepts, then, are Inceptisols that have light-colored surface horizons. The subsurface horizons have been altered and are low in minerals. They are used for woodland and range in Alaska and northwestern United States (scattered areas in the northern and middle Rocky Mountains). In Kansas and Oklahoma they are used as pastureland, for silage corn, and small grains. In the northeastern United States Ochrepts are used as hay lands, the largest area being on the Appalachian Plateau from New York to Tennessee and northern Alabama.

Cryochrepts (Subarctic Brown Forest soils). These are Ochrepts of cold climates as the term implies.

Dystrochrepts (Sols Bruns Acides and some Brown Podzolic and Gray-Brown Podzolic soils). These are rather infertile soils. They are moist, low in bases, and have no free carbonates in the B horizon.

Eutrochrepts (Brown Forest Soils). These are fertile. They are moist and are either high in bases, have free carbonates in the subsurface horizons, or both.

Fragiochrepts (Sols Bruns Acides and some Brown Podzolic and Gray-Brown Podzolic soils, all with fragipans). These soils have a dense, brittle, but not indurated horizon in the subsoil.

Ustochrepts (Reddish Chestnut Soils) are found in the Steppes where they have developed under scanty precipitation. The precipitation is often lacking in the summer season for long periods. Some former Lithosols of less than 20 inches in depth are included in this group.

Xerochrepts (formerly Regosols). The Xerochrepts have formed under a Dry-Summer Subtropical climate in which the precipitation is a wintertime phenomenon. In the warm season the soils are continually dry for lengthy periods.

Tropepts

The prefix *trop* for "of the solstice" is indicative of the existence of continually warm conditions. These are the former Latosols of the Tropical Wet and Dry climates. They are used for pineapple and irrigated sugarcane in Hawaii.

Umbrepts

The prefix *umbr* is from the Latin for "shade" or dark colored. The Umbrepts have thick dark-colored surface horizons that are high in organic matter and B horizons that are low in bases. They exist under a wide range of climatic conditions, and are confined largely to the Cascade Mountains of Oregon and the Coastal Ranges of the Pacific Northwest.

Cryumbrepts (Tundra soils). These are Umbrepts of cold climates at high altitude.

Haplumbrepts (Brown Forest soils). These Umbrepts are found in temperate to warm regions.

Xerumbrepts (Regosols). The Xerumbrepts are found in summer-dry climates. During the warm season the soils are dry for long periods.

mollisols

The Mollisols have nearly black friable surface horizons that are high in organic matter and bases. They have formed for the most part in the subhumid and semiarid margins of the Humid Continental and Humid Subtropical climates.

Albolls

The prefix *alb*, Latin for "white" is indicative of the bleached eluvial horizon that is characteristic of Albolls. They are found in flat terrain and high closed depressions. They have a seasonal perched water table. A nearly black surface horizon is underlain by a bleached, mottled horizon which, in turn, lies over a mottled and gray horizon of clay accumulation. Albolls are found in small scattered areas and are used for small grains, hay, pasture, and range.

Argialbolls (Planosols and Soloths). These have a B horizon of clay accumulation and contain no sodium.

Aquolls

The prefix *aqua*, Latin for "water" indicates the wet character of these Mollisols. They are seasonally wet, and have a thick, nearly black A horizon. The B horizon is gray. Major areas of Aquolls are the middle Mississippi River valley, lower Ohio and Missouri valleys, coastal Texas and Louisiana. One other area of note is the glacial lake plain of North Dakota and Minnesota. They are used for pasture and, where drained, for small grains, corn, and potatoes in the North-Central States, and rice and sugarcane in Texas and Louisiana.

Argiaquolls (Humic-Gley soils) are Aquolls whose B horizon contains considerable clay.

Calciaquolls (Solonchaks). They have large amounts of calcium carbonate accumulations throughout the profile.

Cryaquolls (Alpine Meadow soils). These are Aquolls formed in high altitudes under Tundra-like, alpine climates.

Duraquolls (Humic-Gley soils with hardpan). The hardpan in these soils is quite hard since it is cemented with silica.

Haplaquolls (Humic-Gley soils). These are Aquolls in which mineral materials have been altered or removed from the profile and in which there has been no calcium carbonate or clay accumulation.

Borolls

The prefix, *bor*, comes from the Greek meaning "cold," hence the Borolls are Mollisols formed in cool and cold climates of the continental interior. Most have a black A horizon of some depth. The largest areas of Borolls are in Minnesota, North Dakota and Montana.

They are used for small grains, hay and pasture in the North Central states and for rangeland, woodland, and some small grains farther west.

Argiborolls (Chernozems). These have developed in the cool climate of the subhumid to semiarid margin of the Humid Continental climates. The dark surface horizon is underlain by a B horizon of clay accumulation.

Cryoborolls (Chernozems). These are high latitude soils of the subhumid margin of the Humid Continental climates.

Haploborolls (Chernozems) are similar to the above but contain no horizon of clay accumulation.

Natriborolls (Solonetz soils). These Borolls do have a surface horizon of clay accumulation as well as sodium (alkali) accumulation. These are relatively small inclusions in the other Borolls.

Rendolls

The term is a modification of the older term Rendzina, consequently these Mollisols have subsurface horizons containing large amounts of calcium carbonate but no accumulation of clay. They have been used for cotton, corn, small grains, and pasture.

Udolls

The prefix *ud* is from the Latin for "humid." The Udolls are also found on the subhumid margins of the Humid Continental and Humid Subtropical climatic types. They are usually warm and moist and have no accumulation of gypsum nor of calcium carbonate in the B horizon. They have been used for corn, small grains, and soybeans. The largest area of Udolls is in the Corn Belt with an extension into the livestock and cash grain areas of eastern Kansas and Oklahoma.

Argiudolls (Brunizems and Reddish Prairie soils). As the prefix *arg* (from the Latin *argilla* "clay") indicates, these are the Udolls in which considerable clay has accumulated in the B horizon.

Hapludolls (Brunizems and some Regosols, Brown Forest and Alluvial soils). These are Udolls that have horizons from which some materials have been removed or altered but have no subsurface horizon of clay accumulation. Some of the soils of this group are shallower than 20 inches to bedrock.

Ustolls

The prefix *ust* is from the Latin *ustus* "burnt." These Mollisols, mostly in the semiarid or Steppe climates, are found extensively on the Great Plains from South Dakota to Texas. The soils are intermittently dry for long periods or have an accumulation of salts or carbonates in the B horizon. Their major uses are for range, wheat, sorghums and corn, and some irrigated crops.

Argiustolls (Chernozem, Chestnut and some Brown soils). In these soils there is a horizon of clay accumulation. It is relatively thin and brownish in color.

Calciustolls (Calcisols) are soils having considerable calcium carbonate material throughout the profile. They may have a petrocalcic horizon, an indurated layer cemented by carbonates, or a zone of calcium carbonate or gypsum accumulation. Some are former lithosols and are shallow to bedrock.

Haplustolls (Chernozem, Chestnut, and some Brown soils). High base content is found in the B horizon but without large accumulations of clay, calcium carbonate, or gypsum. Some are less than 20 inches to bedrock.

Natrustolls (Solonetz soils). These have a high content of clay and sodium (alkali) in the B horizon.

Xerolls

The Xerolls (from *xeros*, Greek for "dry") are found in the Subtropical Dry Summer and Steppe climate zones. Since precipitation is received during the winter, the soils are continually dry for long periods during the growing season. They are most extensive on the Columbia Plateau of Washington and Oregon and the Great Basin in western Utah. They find their best use for wheat, range, and irrigated crops.

Argixerolls (Brunizems) are Xerolls that have a B horizon of clay accumulation that is relatively thin or is brownish in color.

Calcixerolls (Calcisols). These have a calcareous surface horizon and a B horizon in which large amounts of calcium carbonate or gypsum have accumulated. The calcium carbonate of the B horizon may or may not be cemented.

Durixerolls (Brunizems with a hardpan). In these the hardpan (durapan) is cemented with silica.

Haploxerolls (Chestnut and Brown soils). The subsurface horizon is high in bases but does not have large accumulations of clay, calcium carbonate, or gypsum.

Palexerolls (Brunizems). The surface horizons are of fairly light to medium texture, grading quickly into a B horizon in which a hardpan (petrocalcic horizon) is cemented with carbonates; or it may have a thick layer of clay accumulation, reddish in color.

oxisols

The Oxisols are found in tropical and subtropical climates at low or moderate elevations. In the United States they are found only in Hawaii. The soils are mixtures of kaolin, hydrated oxides, and quartz, and are low in weatherable minerals.

Humox

The prefix *hum* (Latin *humus*) indicates the presence of organic matter. These are Oxisols that are moist all or most of the time. They have a high content of organic matter but are low in bases. They are used for sugarcane, pineapple, and pasture in Hawaii.

Orthox

From the Greek for "true," the prefix *ortho* indicates that the Orthox suborder is the truest or most common of the Oxisols. These are former Lithosols. They are moist all or most of the time, have a moderate or low content of organic matter, and are relatively low in bases. They are used for sugar cane, pineapple, and pasture.

Ustox

The Ustox suborder is found in dry climates where the summers are hot. *Ust,* the prefix, is from the Latin *ustus,* meaning "burnt." These are former Lithosols. In some part of the soils, they are dry for long periods during the year. They are used for pineapple, irrigated sugarcane, and pasture.

spodosols

The Spodosols are soils with a low base supply, and have in the subsurface horizons an accumulation of amorphous materials consisting of organic matter plus compounds of aluminum and iron. They are formed in acid, coarse-textured materials in Humid Continental climatic areas of northern Michigan, Wisconsin, upper New York and New England; and Humid Subtropical coastal Florida.

Aquods

The Aquods (*aqua,* Latin for "water"), are seasonally wet, ranging widely from the Arctic to the Tropical regions. They are largely used for pasture, range, and woodland, and in Florida for citrus and truck crops.

Haplaquods (Ground-Water Podzols). These have a B horizon that contains dispersed aluminum and organic matter but only small amounts of free iron oxides. They are used for woodland and pasture. Some truck crops and citrus are produced where these soils are drained.

Sideraquods (Ground-Water Podzols) are Aquods that have appreciable amounts of free iron in subsurface horizons.

Orthods

These are the most widely distributed of the Spodosols. Orthods have a horizon in which organic matter as well as compounds of iron and aluminum have accumulated. They are used for woodland, hay, pasture, fruit, and on gently sloping areas, potatoes and truck crops.

Cryorthods (Podzols) are less than 20 inches to bedrock and practically all contain a podzolized layer, ashy white in color and highly acid.

Fragiorthods (Podzols and Brown Podzolic soils, both with fragipans). These soils have a dense brittle (but not indurated) horizon or fragipan below a horizon that has accumulations of organic matter and compounds of iron and aluminum.

Haplorthods (Podzols and Brown Podzolic soils). The Haplorthods have no dense, brittle, or indurated horizon or fragipan. They have developed in cool climates and have a B horizon in which organic matter and compounds of iron and aluminum have accumulated.

ultisols

The Ultisols are moist, low in bases, and have subsurface horizons of clay accumulation. They are found largely in Humid Subtropical climatic areas and extend from New Jersey to Texas to Missouri on coastal plain sediments. Although moist for most of the year, some soils are dry part of the time during the warm summer season. All are considered as warm soils in which annual soil temperatures average above 47°.

Aquults

The Aquults as the term implies are wet, at least seasonally. They have mottled horizons, iron-manganese concretions, or gray-colored horizons. They are limited in their use largely to pasture and woodland, and where drained, some hay, cotton, corn, and truck crops are grown.

Fragiaquults (Planosols with a fragipan). These Aquults have a dense, brittle, but not indurated horizon (fragipan).

Ochraquults (Low-Humic Gley soils). The distinguishing feature of these Aquults is a light-colored or thin black surface horizon.

Umbraquults (Humic-Gley soils). These are Aquults that have a thick black surface horizon.

Humults

A high content of organic matter is characteristic of the Humults. They are found in the Marine West Coast climate and in Hawaii within the Tropical Wet and Dry climate. In both, precipitation is high. They are used for woodland and pasture on the steeper slopes and for small grains and truck and seed crops on the more gentle slopes. In Hawaii they are used for sugarcane and pineapple plantations.

Haplohumults (Gray-Brown Podzolic and some Reddish-Brown Latteritic soils). These are Humults found in the temperate West Coast Marine climate. The B horizon may be either one of clay accumulation, with no pan development, one in which there may be considerable weatherable mineral matter, or both.

Tropohumults (Reddish-Brown Lateritic soils). These are found under tropical climates. Like the Haplohumults, the Tropohumults may have a relatively thin B horizon of clay accumulation, or one that contains considerable mineral matter easily weatherable, or both.

Udults

The Udults are formed in humid climates such as the Humid Subtropical in which there are few dry periods. They are relatively low in organic matter in the B horizon. They are used for general farming, woodland and pasture, cotton and tobacco.

Fragiudults (Red-Yellow Podzolic soils with a fragipan). The Fragiudults have a dense brittle but not indurated horizon or fragipan, in or below the B horizon of clay accumulation.

Hapludults (Red-Yellow Podzolic and some Gray-Brown Podzolic soils). The B horizon is one in which clay has accumulated in a relatively thin layer or one in which there is an appreciable content of weatherable minerals, or both. They are extensive on the crystalline rocks of the Piedmont Plateau.

Paleudults (Red-Yellow Podzolic soils). The B horizon is one of thick clay accumulation with few weatherable mineral materials. Extensive areas are found on the coastal plain from Virginia to Texas and northward to Missouri and Kentucky.

Rhodudults (Reddish-Brown Lateritic soils). The B horizon is one of dark-red color and clay accumulation.

Xerults

These are found in the Subtropical Dry Summer climates and are continually dry for long periods during the summer. The B horizon is relatively low in organic matter. They are used for range and woodland.

Haploxerults (Reddish-Brown Lateritic and some Red-Yellow Podzolic soils). The B horizon is one that has a relatively thin zone of clay accumulation, much weatherable mineral matter, or both. The Western Sierra and northern Great Valley of California are the major regions in which the Haploxerults are found.

vertisols

The Vertisols are clay soils that crack deeply when dry and are found largely in the Black Prairies of Mississippi and Texas as well as coastal Texas. They form under climatic conditions having distinct wet and dry periods every year.

Torrerts

Formerly Grumusols, these Vertisols are usually dry, having wide deep cracks that remain open throughout the year. They are used for range and some irrigated crops.

Uderts

Unlike the Torrerts, the Uderts are moist and have cracks that open and close at least once during the year. The cracks remain open for less than two months or intermittently for periods of more than three months. Cotton, corn, small grains, pasture, and some rice are grown on these soils. The largest area is found in coastal Texas.

Chromuderts and *Pelluderts* are former Grumusols. The former have a brownish surface horizon, the latter have a dark gray to black surface horizon.

Usterts

These are Vertisols that have wide, deep cracks that usually open and close more than once during the year and remain open intermittently for three months or more. They are used for general crops and range plus some irrigated cotton, corn, citrus, and truck crops in the Rio Grande valley.

Chromusterts and *Pellusterts* (former Grumusols) have a brownish A horizon in the first instance and a dark gray to black A horizon in the second. The parent materials are heavy clays.

Xererts

These Vertisols have wide, deep cracks that open and close once each year and remain open continuously for more than two months. They are used for irrigated small grains, hay and pasture.

TABLE 5.7
Orders, Suborders, and Great Group with respect to major aspects of Climate.

WARM SOILS

Mean Annual Soil Temperature Higher Than About 47°F

Order	Moist	Wet

Alfisols	Udalfs
	  Fragiudalfs (G.-B. Podzolic with fragipan)
	  Hapludalfs (g-B podzolic without fragipan)
	  Paleudalfs (R-Y and G-B podzolic soils)
		Aqualfs
		  Albaqualfs (Planosols)
		  Fragiaqualfs (Low-Himic Gley with fragipan)
		  Glossaqualfs (Planosols)
		  Natraqualfs (Solonetz)
		  Ochraqualfs (Low-Humic Gley)

Aridisols		

Entisols	Psamments
	  Quartzipsamments (Regosols)
	  Udipsamments (Regosols)
		Aquents
		  Haplaquents (Low-Humic Gley)
		  Hydraquents
		  Psammaquents (Low-Humic Gley, Regosols)

Histosols		Histosols (Moderately to highly decomposed O.M.)

Inceptisols	Andepts
	  Dystrandepts (Ando soils)
	Ochrepts
	  Dystrocrepts (Sols Bruns Acides, Gr.-B. Podzolic B. Podzolic)
	  Eutrocrepts (Brown Forest)
	  Fragiocrepts (Sols Bruns Acides, Br. and G-B Podzolic)
	Tropepts (Latosols)
	Umbrepts
	  Haplumbrepts (Brown Forest Soils)
		Aquepts
		  Haplaquepts (Low-Humic Gley)
		  Humaquepts (Humic Gley)

Mollisols	Udolls
	  Argiudolls (Brunizem, Reddish Prairie)
	  Hapludolls (Brunizem, Br. Forest, Alluvial Regosol,Lithosols)
		Aquolls
		  Argiaquolls (Humic Gley)
		  Calciaquolls (Solonchaks)
		  Haplaquolls (Humic Gley)
		  Duraquolls (Humic Gley with hardpan)

Oxisols	Orthox (Latosols)	Humox (Latosols)

Spodosols		Aquods
		  Haplaquods (Ground-water Podzol)
		  Sideraquods (Ground-water Podzol)

Ultisols	Humults
	  Haplohumults (Red-Br. Laterite)
	  Tropohumults (Red-Br. Laterite)
	Udults
	  Fragiudults (Red-Yellow Podzolic w fragipan)
	  Hapludults (Red-Yellow, G-B Podzolic)
	  Paleudults (Red-Yellow Podzolic)
	  Rhodults (Reddish-Brown Laterites)
		Aquults
		  Fragiaquults (Planosols w. fragipan)
		  Ochraquults (Low-Humic Gley)

Vertisols	Uderts
	  Chromuderts (Grumusols)
	  Pelluderts (Grumusols)	

Misc. Land Types		

COOL SOILS

Mean Annual Soil Temperature Lower Than About 47°F

Dry	Moist	Wet
alfs aplustalfs (Reddish Chestnut and Reddish Brown soils) aleustalfs alfs urixeralfs (noncalcic Br. with hardpan) aploxeralfs (noncalcic Br. soils) alexeralfs (noncalcic Br. soils)	Boralfs Cryoboralfs (Gr. Wooded soils) Eutroboralfs (Gr. Wooded soils) Fragiboralfs (Gr. Wooded with fragipan) Glossoboralfs (Gr. Wooded soils)	
gids urargids (Desert, Red Desert, Sierozem and some Br. w. aplargids (Sierozem, Desert, Red Desert and some Br. soils hardpan) adurargids (Solonetz with hardpan) atrargids (Solonetz) aleargids (Sierozem, Desert and Red Desert soils) hids alciorthids (Calcisols, Lithosols) amborthids (Sierozem, Desert and Red Desert soils) urorthids (Regosols, Calcisols, Alluvial, w. hardpan) aleorthids (Calcisols) alorthids (Solonchaks)		
vents orrifluvents (Alluvial) diifluvents (Alluvial) erofluvents (Alluvial) thents orriorthents (Regosols, Lithosols) storthents (Regosols, Lithosols) erorthents (Regosols, Brown, Alluvial) mments orripsamments (Regosols) stipsamments (Regosols) eropsamments (Regosols)	Orthents Cryorthents (Alluvial, Regosols, Lithosols) Psamments Cryopsamments (Regosols)	
		Histosols—Peat
depts Eutrandepts (Brown and Reddish Brown soils) hrepts stocrepts (Lithosols)	Andepts Cryandepts (Brown and G.B. Podzols) Ocrepts Cryocrepts Subarctic Br. Forest) Umbrepts Cryumbrepts (Tundra soils)	Aquepts Cryaquepts (Tundra soils)
tolls Argiustolls (Chernozem, Chestnut, Brown) Calciustolls (Calcisols, Lithosols) Haplustolls (Chernozem, Chestnut, Brown, Lithosols) Natrustolls (Sononetz) rolls Argixerolls (Brunizems) Calcixerolls (Calcisols) Durixerolls (Brunizems with hardpan) Haploxerolls (Chestnut, Brown) Palexerolls (Brunizem with hardpan)	Borolls Arigiborolls (Chernozem) Cryoborolls (Chernozem) Haploborolls (Chernozem) Natriborolls (Solonetz) Aquolls Cryaquolls (Alpine Meadow soils)	Aquolls Cryaquolls (Alpine Meadow soils)
tox (Latosols)	Orthods Cryorthods (Podzols) Fragiorthods (Podzols and Br. Podzols with fragipans) Haplorthods (Podzols and Br. Podzols)	
erults Haploxerults (Red-Br. Laterites, R.Y. Podzol)		
sterts Chromusterts (Grumusols) Pellusterts (Grumusols) ererts Chromoxererts (Grumusols) Pelloxererts (Grumusols)		

Chromoxererts (Grumusols) are characterized by a brownish A horizon. *Pelloxererts* (also former Grumusols) are distinguished from the Chromoxererts by a dark-gray to black surface horizon. Both have developed upon clayey parent materials.

Table 5.7 lists in tabular form, Orders, Suborders, and Great Groups with respect to major aspects of climate. The terminology of the old classification is added for comparison purposes.

SELECTED REFERENCES

MAUSEL, PAUL W. "A Selected Application of the 7th Approximation in Soils Geography," *The Professional Geographer*, March, 1968, pp. 116-122.

SIMONSON, ROY W. "Soil Classification in the United States," *Science*, September 28, 1962, pp. 1027-1034. A concise review of attempts at classification but more important is a discussion of the 7th Approximation.

U.S. Soil Conservation Service, Soil Survey Staff. *Soil Classification: A Comprehensive System, 7th Approximation.* Washington: 1960.

————. *Supplement to Soil Classification System (7th Approximation)*, Washington: 1967.

U.S. Department of the Interior, Geological Survey, National Atlas sheet number 86, 1967. Distribution of Principal Kinds of Soils, Orders, Suborders and Great Groups.

U.S. Department of Agriculture. "Soil," *Yearbook of Agriculture*. Washington: 1957. The last part of the book is a discussion of agricultural land use in the United States by regions. Soils characteristics (old Classification terminology) are discussed.

Man's Use of the Soil

Without question the soil is of great importance to man. He obtains much of his food crops directly from tillage of the soil or from livestock fattened on feed grains, silage, hay, or from pasture and grazing. He builds his roads and highways upon the soil. He must utilize the soil for the site of his office buildings, his factories, his shopping facilities, his homes, and his parking lots. Each use presents a different set of soil problems that must be solved. In an attempt to forestall certain of these problems man has classified the soils as to their potential for the use to which he wishes to put them. A few are noted below.

Classification of Land for Agricultural Purposes

The past 75 years have yielded much information on soil and water management in relation to lands, and various workers have offered land classifications (basic to which are soil types and their physical properties) as to their utility and productivity for crops. Using the research and soil surveys they have been able to estimate acre yields of grain crops and feed crops as well as pasture yields in the fattening of livestock. On the basis of soil types and under various levels of management, similar predictions of crop yields have been attempted quite successfully, both on irrigated and non-irrigated lands of similar soil type.

Such classification schemes (which take into consideration basic information supplied by the published reports of the Soil Conservation Service) have much broader uses. Not only is the soil type, with all its physical and chemical properties noted insofar as crop production is concerned, but also such characteristics as susceptibility to erosion by water and wind, ability to absorb rainfall rapidly (permeability), depth to bedrock, and so forth. The steeper, longer slopes allow for rapid run-off and subsequent erosion of topsoils, as does the poor practice of planting row crops down slope, that is, across contours rather than around the slope on the contour. Muck and peat soils once drained are readily susceptible to fire and wind erosion, and attempting to grow deep-rooted crops such as alfalfa or forest crops on soils shallow to bedrock is an obvious impossibility (Figure 6.1).

These sorts of contingencies are considered in a land classification along with the soil types themselves, to give a more complete economic land cassification. Although this is of a temporal nature, it is very useful to man in the present state of technology.

One of these classifications in wide use for agricultural land takes conservation practices into consideration and thereby attempts to put the land to its best use for the longest period of time. Instead of delineation of soil type boundaries on maps, eight land-capability classes with sub-classes are recorded (Figure 6.2). Classes I through IV are considered suitable for profitable and extended cultivation when specific management practices are followed. Class I land requires little if any conservation practices to obtain high level production of crops, pasture, range, or woodland. Classes II, III, and IV are progressively less desirable. More limitations to their use exist, or cost for maintenence of soil protection or improvement in productivity is increased. Land Classes V, VI, and VII, while considered unsuitable for cultivation, are useful for grazing and forestry. The last, Class VIII, is unsuitable for any use except for wildlife, recreation, and watershed uses. Sub-classes in the system provide further information as to the other characteristics that make the uses of the land in each class more restrictive.

The above system is largely restricted to agricultural uses of land in humid areas. Where climate is more hazardous and agricultural practices include the addition of irrigation water to the land, the suitability of similar soil types for a specific crop use may change drastically.

The Bureau of Reclamation devised a system into which certain economic considerations were incorporated. Of the six classes, the first four are suitable for irrigation agriculture. Class 1 lands are those on which topography, drainage and soil characteristics are very favorable to crop production and have the potential to repay costs of irrigation. Class 2, under irrigation practices, are intermediate in their subsequent ability to repay irrigation costs due to higher development and

Figure 6.1. Soils shallow to bedrock are generally of limited use. They may best be used for pasture or grazing land or a source of stone for construction purposes. Courtesy U.S.D.A.

Figure 6.2. The land-capability classes are here illustrated upon a landscape. As agricultural land, Class I is the best. The choices become less and the risks increase as the class number increases so that Class VIII land is unsuitable for any use except for wildlife, recreation, and watershed uses. Soil Conservation Service.

drainage costs, or in all respects are similar to Class 1 except that the soil is of a lesser quality and hence likely to be a poorer producer. Class 3 lands are border-line soils. The deficiencies of slope, soil, or drainage characteristics are greater than in Class 2. Irrigation canals built in excessively permeable soils may have to be lined (Figure 6.3). Class 4 lands require further detailed study of an economic and engineering nature to determine their feasibility for irrigation. They may be isolated, requiring costly construction to bring water to them, or may require special treatment to obtain a profitable crop return.

Class 5 lands are those which now are unsuitable for irrigation, even though otherwise suitable from a soils point of view. It may be that irrigation water cannot be brought to them cheaply enough at present to return a profit. They may be too small in area or they may be isolated. If it is determined that this can be overcome by some technological change or procedure, it will then be upgraded. If not, it will be discarded and put into Class 6 land. Class 6 lands are those that are so poor in soils, topography or drainage characteristics that it would be unprofitable to irrigate under any circumstances. Such land may be above the canal system. Pumping expenses to lift water to this land would be exorbitant or the topography may be so rough that the engineering costs to bring water to the land could not be repaid. They may be excessively stony or irrigation water may percolate through them too rapidly, or the other extreme may exist where the permeability is so poor that water may move through the soil so slowly that standing water remains on the surface indefinitely.

An example of the symbolization used in this classification of the Bureau of Reclamation is as follows:

Land Class	Soils	Topography	Drainage
3	S	T	D
M	H	A	J

The land Class 3 is so designated because of some soils, topography, and drainage deficiencies indicated by the "S", "T", and "D". The soils deficiency is indicated by the "H" in the denominator under "S". The "M" and the "H" are soils textural symbols, indicating a medium surface texture and a heavy subsurface texture. A salinity problem is also found as designated by the symbol "A" for alkali in the denominator. The topography and the drainage of this area are not entirely satisfactory as indicated by the "T" and the "D" symbols. For Class 3 land the topography is such that slopes of 5 to 8 percent are the rule and "runs" are short but not less than 150 feet. The topography factor

Figure 6.3. Irrigation canals built in excessively permeable soils may have to be lined to reduce the water loss. Concrete has been sprayed into the irrigation ditches as is being done in this photo. Asphalt, plastic, and other materials have also been used. Courtesy U.S.D.A.

"J" indicates that the field pattern is irregular, following the general irregular character of the surface. Other symbols are listed and defined in Table 6.1.

The classification of forest and grazing lands has also been attempted so that not only an inventory of types and species of trees and commercial and non-commercial forest and grazing lands was available, but productivity, depletion, and replenishment information for a sustained yield of both was indicated.

Man has by such classifications been able to identify his land resources as to their present and potential productivity. Each classification has made available to him information which enables him to so modify the use of that land as to obtain the best possible return. What is true at present may not be true, of course, should the technology improve in the future. Nevertheless, the classifications are not fixed or rigid, and may and will be modified to serve the men that made them.

TABLE 6.1

Land Classification Scheme of the Bureau of Reclamation.

LAND CHARACTERISTICS	Symbols		CLASS 1—Arable
	Basic subclass	Inform. & Defic.	
SOILS s			
Textures			Sandy loams to clay loams.
Very light		v	
Moderately light		l	Sandy loams and loams.
Moderately heavy		m	Silt loams and clay loams.
Heavy		h	
Depth			
To incoherent sands, clean gravel or cobbles.		k	36" or more of good free working soil of fine sandy loam or heavier, or 42" of sandy loam. If gravel contains soil, depth may be minimum of 30" depending upon the water-holding capacity.
To solid sandstone, shale bedrock, raw wash from shale impervious clay substrata, or any other impervious substrata.		b	60" plus or 56" with minimum of 6" of gravel overlying impervious material or sandy loam throughout.
To highly concentrated (gray to white) penetrable lime zone.		s	18" plus with 60" penetrable.
Alkalinity and Salinity		a	Total salts* not to exceed .2% or 4 millimhos per cm. May be higher under good leaching and drainage conditions. pH 8.6* or less on paste. May go to 8.8 below 42". pH 9.0 or less on 1-5 dilution, associated with good permeability.
			*See Supplement
TOPOGRAPHY t			
Slope		g	0.15% to 25% incl.
*Irrigation pattern		j	400 fee minimum run. 8 acres minimum size.
Surface (leveling)		u	Light 0-175 cuvic yards excavation per acre. 0.0 to 0.22' average cut and fill.
*Cover (vegetation)		c	
Low brush (6 ft. and less)			0% to 100% cover.
High brush (over 6 ft.)			0% to 50% cover.
Trees (6 to 15" diameter)			0 to 8 trees per acre.
Stones and cobbles		r	0 to 10 cubic yards per acre.
DRAINAGE d			No drainage anticipated.
MISCELLANEOUS			
High areas above level of irrigation	h		Occurs only in classes 5 and 6. Irrigability in question when used with class 5 and definitely non-irrigable when used with class 6.
Isolated areas	i		Occurs only in classes 5 and 6. To be shown same as high areas as $\frac{6h\ (2s)}{mmk}$ or $\frac{6i\ (3t)}{mmu}$.

CLASS 4— Limited arable. Includes lands which fail to meet the minimum requirements of Class 3 under irrigation, such as pasture, suburban use, etc. (See handbook 2.2.5, 2.4.2H-5,

CLASS 5— Non-arable. Includes lands which require additional economic and engineering studies of corrective works and reclamation through application of these works. Upon complesymbols listed above for appropriate classification symbols.)

CLASS 6— Non-arable. Includes lands which do not meet the minimum requirements of the next non-arable land. (See handbook paragraphs 2.4.2B (2-6), 2.5.5A and symbols listed

*Note: Irrigation pattern—Minimum are fields that are approximately rectangular. As shapes become more irregular minimum allowable rises shall be increased proportionally. A field is defined as a body of land lying in approximately the same plane, and feasible of being irrigated with single layout of parallel border dikes or parallel corrugations.

CLASS 2—Arable	CLASS 3—Arable
Loamy sands to very permeable clays. Loamy sands or gravel sufficient to moderately reduce productivity and moisture-holding capacity. Sandy loams and loams. Silt loams and clay loams. Very permeable clay.	Loamy sands to permeable clays. Loamy sands or gravel sufficient to markedly reduce productivity and moisture-holding capacity. Sandy loams and loams. Silt loams and clay loams. Permeable clay.
24" plus of good free working soil of fine sandy loam or heavier, or 30" of loamy sand. If gravel contains soil, depth may be minimum of 20" depending upon the water-holding capacity. 48" plus or 42" with minimum of 6" of gravel overlying impervious material or loamy sand throughout.	18" plus of good free working soil of sandy loam or heavier, or 24" of loamy sand. If gravel contains soil, depth may be minimum of 12" depending upon water-holding capacity. 42" plus or 36" with minimum of 6" of gravel overlying impervious material or loamy sand throughout.
14" plus with 48" penetrable.	8" plus with 36" penetrable.
Total salts* not to exceed .4% or 8 millimhos per cm. May be higher under good leaching and drainage conditions. pH 8.8 or less on paste. pH 9.3* or less on 1-5 dilution. pH spread between paste and 1-5 less than 1:1 if pH on 1-5 dilution is 9.0 or more. *See Supplement	Total salts* not to exceed .6% or 15 millimhos per cm. May be higher under good leaching and drainage conditions. pH 8.8 or less on paste. pH 9.6* or less on 1-5 dilution. May be higher below 42". pH spread between paste and 1-5 less than 1:1, if pH on the 1-5 dilution is 9.0 or more. *See Supplement
0.0% to 0.14% and 2% to 5%. 300 feet minimum run. 5 acres minimum size. Medium 175 to 350 cubic yards excavation per acre. 0.22 to 0.44' average cut and fill.	5% to 8%. 150 feet minimum run. 2 acres minimum size. Heavy 350 to 650 cubic yards excavation per acre. 0.44 to 0.81' average cut and fill.
Not applicable. 50% to 100% cover. 8 to 18 trees per acre. 10 to 22 cubic yards per acre.	Not applicable. Not applicable. 18 to 35 trees per acre. 22 to 45 cubic yards per acre.
Slight drainage problem anticipated but may be improved at relatively low cost.	Drainage problem anticipated but may be improved by expensive but feasible measures

but which special economic and engineering studies have shown capable of feasible restricted use
2.5.5a, and symbols listed above for appropriate classification symbols.)
to determine their irrigability and lands classified as temporarily non-productive pending construction
tion of studies, these lands are placed in proper class. (See handbook 2.4.2H-6 and 2.5.5A and

higher class mapped in a particular survey and small areas of arable land lying within larger bodies of
above for appropriate classification symbols.)

*Note: Cover—Cover is defined as shaded area. The class shall be lowered proportionately as different
types of cover occur in combination. As the size of trees increases over 15" diameter, the
allowable number per acre shall be decreased.

County Soil Survey Reports

Although the Soil Survey report is not intended to eliminate onsite sampling, a Soils Survey report can find good use in the following areas as well as in agriculture:

1. To make soil and land use studies to aid in the selection and development of sites for industrial, business, residential, and recreational activities.
2. Preliminary estimates of those properties of soils that will aid in the planning of irrigation and drainage systems, waterways and diversion terraces, ponds and lakes.
3. Estimation of ground conditions for selection of location of highways, airports, and to assist in those areas to be examined in greater detail.
4. Location of gravel, sand, and stone of suitable quality for construction purposes, and those soil types suitable for movement of heavy vehicles.
5. Correlation of soil type with engineering structures as an aid to proper design and maintenance of such structures.
6. Supply supplementary information especially if the map is constructed on a photograph of good scale.

Classification of Land for Urban and Suburban Development

With something around 92 per cent of the U.S. labor force engaged in activities not related to the farm but concentrated in city areas instead, urban and suburban development has raced ahead, swallowing up farm lands, both good and bad, at a rapid rate. To most land developers and builders all farm land was the same, soil was soil no matter where it was and as such was the potential site for a new building. All over the country land was changed from its former often varied agricultural use to a monotonous pattern of rectangular, treeless home sites. During all this development, misuse and abuse of soils through disregard of their physical properties caused many disasters requiring many expensive repairs. These costs might have been avoided had due regard been paid to the soil capability.

Septic tank filter fields did not function properly where permeability of soils was poor (Figure 6.4). Flooding on lowlands and terraces inundated many houses, cracks developed in foundations often before the house was completed, foundations or even whole developments slipped down slopes (Figure 6.5). All costly mistakes, some-

Figure 6.4. Where permeability of soils is poor, trouble with septic tanks is certain. The illustration shows effluent surfacing in a septic tank filter field in soil that is incapable of absorbing great quantities of water. The problem is especially bad during wet weather. Photo Soil Conservation Service.

times complete losses—these occurences could have been avoided had proper use been made of the wealth of information packed into the soil survey bulletins. Should no such bulletin be available, the soils specialist can quickly make a survey for the area prior to its development.

To most, expert and layman alike, many soils are similar on the surface. Plant cover appears to be quite uniform, giving no hint as to what lies beneath the surface. The above mentioned pitfalls can be avoided if the characteristics of each soil layer are interpreted by a specialist beforehand. Soils vary greatly within short distances even though the topography may be flat, and very definitely when the land is undulating or rolling and transected by streams and stream terraces. The soil specialist can aid planners and builders by pointing out which areas are unsuitable for structures of a specific type and

Figure 6.5. Unstable soil, whether on lowland or on a slope, is unsuitable for construction of many types. Slides occur when the soil is saturated and damage such as illustrated will result. Soil survey reports are extremely helpful in preventing use of land of this kind for home construction. Photos Soil Conservation Service.

aid in the design and construction of buildings and roads in other areas. With such help, areas of flood hazard will be avoided or developed in such a manner as to enhance surrounding sites by the possible incorporation of recreational areas of one type or another into the development. Even in its simplest form this would eliminate the use of the prevalent rectangular block in home site developments. Fortunately this adherence to the rectangular block is changing. If in the planning of an urban development, an evaluation of the Soil Survey is made with due regard to the capabilities of the land, the result will be a much more pleasant urban environment. Blending the urban with the physical landscape will utilize the latter to its fullest for what it is best suited, whether building site, park, road, lake, wooded area, or playground (Figure 6.6).

The soil survey supplies information on the soil's permeability to water, its susceptibility to flooding, its shear strength and bearing strength characteristics (Figure 6.7), its erosion and corrosion potentials (Figure 6.8), and its possible reaction to frost action as well as its agricultural potentialities. The soil permeability characteristics are readily observable. A field where water stands for days after a shower

Figure 6.6. In planning for new urban development the use of the information from the soil survey will enable making the best use of the soils whether for building site, park, road, lake, wooded area, or playground. This is the planned community of Greenbelt, Md. Courtesy U.S.D.A.

Figure 6.7. The upper photo is illustrative of what has occurred to roads where the soil materials upon which they have been built are unable to support additional weight. The lower photo is of a house built on a concrete slab on piling driven into peat soils. In a subdivision of 20 houses the soil has settled two feet in two years exposing utility pipes and wires. Soil Conservation Service.

and that seldom produces a crop is one of slow permeability. On the other hand, a soil from which water drains away almost immediately after a rain is of moderately rapid to rapid permeability (Table 6.2).

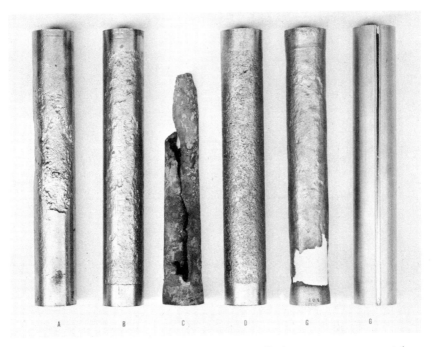

Figure 6.8. The corrosive action of some soils is very great on metal pipes in the soil. Such information available before pipes are laid can forestall costly replacement and accidents.

TABLE 6.2

Soil Permeability.

Very slow	<0.05 inches per hour
slow	$0.05-0.20$ inches per hour
moderately slow	$0.20-0.80$ inches per hour
moderate	$0.80-2.50$ inches per hour
moderately rapid	$2.50-5.00$ inches per hour
rapid	$5.00-10.00$ inches per hour
very rapid	greater than 10.00 in. per hr.

TABLE 6.3

Evaluations of Soil Associations of Seneca County, Ohio.

Some Important Characteristics of Major Soils* Seneca County, Ohio

Soil Association Name & Number	Slope Range	Soil Features and Limitations for Selected Uses							
		Truck Crops (1)	Irrigation (1)	Ponds	Septic Tank Leach Fields	Homesites (2)	Roads & Parking Lots	Pipelines & Sewers	Recreation (3)
Eel-Shoals-Sloan-Genesee 1.	nearly level	few limitations, flooding, wetness	high infiltration, moderate water holding capacity	flooding, seepage	flooding, wetness	flooding, wetness	flooding, wetness	few limitations, pipes may corrode	flooding
Haskins-Haney-Belmore 2.	nearly level to sloping	few limitations, drouthy, temp. wetness	high infiltration, moderate water holding capacity	moderate to high seepage	few limitations on Haney & Belmore; Haskins wet	few limitations on Haney & Belmore; Haskins wet	frost heave, Haskins wet	ditch walls unstable, pipes may corrode	few limitations, Haskins wet
Tawas-Willette 3.	level	few limitations, wetness	high infiltration low to high water holding capacity	moderate to high seepage	wetness	wetness	poor stability wetness	ditch walls unstable, pipes may corrode	wetness
Blount-Pewamo 4.	nearly level to gently sloping	wetness	moderate to moderately slow infiltration, moderate to high water holding capacity	few limitations	wetness, slow permeability	wetness	wetness	pipes may corrode	wetness
Morley-Pewamo 5.	nearly level to gently sloping	wetness, temporary wetness	moderate infiltration, moderate to high water holding capacity	moderate to high seepage	moderately slow to slow permeability, wetness	few limitations on Morley; Pewamo wet	frost heave, Pewamo wet	pipes may corrode	few limitations, on Morley; Pewamo wet
Morley-Blount 6.	gently sloping to steep	erodible on slopes, wetness	moderate to moderately slow infiltration, moderate to high water holding capacity	few limitations	wetness, moderately slow to slow permeability	few limitations, on Morley; Blount wet	frost heave, Blount wet	pipes may corrode	some steep slopes on Morley; Blount wet
Hoytville 7.	nearly level	wetness	moderate infiltration, moderate to high water holding capacity	few limitations	wetness, slow permeability	wetness	wetness	pipes may corrode	wetness
Millsdale-Randolph-Romeo 8.	nearly level to gently sloping	shallow to bedrock, wetness	shallow to bedrock, moderate infiltration & water holding capacity	shallow to bedrock	shallow to bedrock, wetness	shallow to bedrock, wetness	shallow to bedrock, wetness	shallow to bedrock	wetness
Bennington-Pewamo-Cardington 9.	nearly level	wetness, erodible on slopes	moderate infiltration, high water holding capacity	few limitations	wetness, slow to moderately slow permeability	few limitations, on Cardington; Bennington & Pewamo wet	frost heave, Bennington & Pewamo wet	pipes may corrode	few limitations, on Cardington; Bennington & Pewamo wet

* Soil characteristics in this table are on an association rather than an individual soil basis.

(1) These ratings are based on little or no erosion; ratings for truck crops and irrigation apply to soils on nearly level to greatly sloping topography

Table 6.3 is adapted from the general soil map of Seneca County, Ohio. In it evaluations of soil associations with respect to the above uses are made. For example:

1. The Morley-Pewamo soils series have moderate to high seepage, high rate of water percolation, and are of limited use for ponds and lakes.
2. The Eel-Shoals-Sloan-Genessee association, being bottomland soils, are susceptible to flooding and should be used only in very limited ways.
3. The Tawas-Willette (muck soils) have poor stability, are wet and therefore poor for roads, parking lots or home sites.
4. The Morley-Blount soils tend in part to be steep topographically and subject to erosion; therefore if cleared would require the implementation of some erosion control techniques.
5. Many of the soils listed may corrode pipelines and are subject to frost action, necessitating special control measures.

Classification of Land for Tax Purposes

Another facet of man's doing involves taxation and here, too, the soils map can be extremely useful as a basis for equitable tax assessments. Early assessments for tax purposes on agricultural lands were often made more by acreage than value of the land itself. More progressive districts are using those estimates of soil quality as found in modern soil surveys along with estimates of crop yields as a means of spreading the tax load more equitably.

Soil Modification by Man

Soils have been "reconstituted" by the use of agricultural practices which have greatly modified the original soil characteristics and properties. These methods of soil tillage have often been techniques developed in answer to a specific need. Consequently the use of specialized power tools to break up restrictive hardpans, to drain unproductive swamps, to reclaim portions of the continental shelf (as with the Dutch polders for example), to shape irregularly shaped land areas for irrigation use and water control, has put much more land to productive use. Water control is also the means whereby the proper aeration of the soil, so necessary for most microorganisms, is maintained through proper management practice.

Often soils in the traditionally heavy crop-producing regions require replenishment of mineral and organic materials to continue in use efficiently. Toward that end commercial fertilizers and organic residues are supplied to the soils in quantities needed by the specific soil, utilizing improved techniques and machinery. Commercial fertilizers are available with varying amounts of nitrogen, phosphorus and potassium so that the right combination may be obtained for the specific soil. Also the minor or trace elements, very small amounts of which are tremendously beneficial in increasing crops both in quantity and quality, are added where such has been found deficient or lacking by soil analyses. Fertilizer application techniques have greatly changed over the years to the point where aircraft are sometimes employed today.

Procedures are in use to control wind and water erosion, thus stemming soil movement. Many of these involve shaping the land surface and/or use of certain crops to slow velocity of both wind and water (Figure 6.9).

Soil temperature is critical for germination. To some extent soil temperatures are modified by methods that increase or decrease retention of the sun's energy as needed. Manipulation of soil water content will cause an increase or decrease of soil temperatures at will.

Figure 6.9. Reshaping the land to slow the movement of water and wind reduces erosion of the soil. In this case, contour tilling and irrigating on the contour is very helpful in reducing soil erosion.

Agricultural soils, then, have undergone changes from the moment man put them to the plow. What organic matter may have been in the soil was oxidized more rapidly as the soil was turned over, loosened, and mixed with air. With its protective cover of vegetation removed or replaced by cultivated crops, erosion by both wind and water effectively removed great quantities of top soil. In some instances the whole soil was removed and the parent materials exposed to the surface. (Figure 6.10 indicates the extent of soil erosion in the United States. Very few areas have escaped some erosion following the plow and cultivation.)

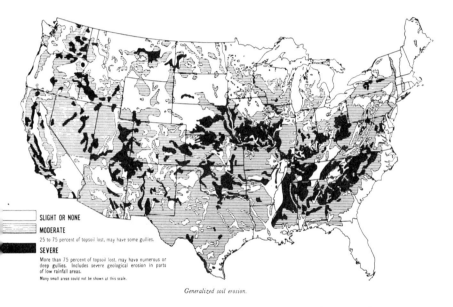

SLIGHT OR NONE

MODERATE
25 to 75 percent of topsoil lost, may have some gullies.

SEVERE
More than 75 percent of topsoil lost, may have numerous or deep gullies. Includes severe geological erosion in parts of low rainfall areas.
Many small areas could not be shown at this scale.

Generalized soil erosion.

Figure 6.10. Generalized Soil Erosion map of the United States. Courtesy U.S.D.A.

The eroded materials carried downstream were deposited in the bottomlands, have filled reservoirs (Figure 6.11) and have covered roads and buildings. At best, the floodplains of the rivers involved became the recipient of the top soils from the lands upstream. If severe erosion had occurred upstream, or had continued for many years, the bottomlands were ruined by being covered with infertile and stony materials. In the one case the land became more fertile, while in the other the land became worthless.

Figure 6.11. The storage capacity of the reservoir is lost as the area behind the dam is filled with sediments. The sediments most frequently are the topsoils from the agricultural lands of the drainage basin which in this case was 462 square miles. The original depth of the reservoir was about 35 feet and it had a capacity of 320 acre feet. The reservoir was abandoned in 1952 after 30 years of use Courtesy U.S.D.A.

Management practices of the soils have not been all bad. The progressive farmer in tune with the times rotated his crops, applied commercial fertilizers at the recommended rates, stabilized his soil against the ravages of wind and water, and returned crop residues to the soil. The soil in time was modified considerably from its natural state and may have become more fertile. The addition of a complete fertilizer to the soil increases its nitrogen, phosphorus, and potassium content, along with other minor yet essential elements. Many unproductive soils in arid lands have been made very productive when water was brought to them. The quality and quantity of water has been effectively increased by reduction of seepage and evaporation losses, removal of water-wasting vegetation near the irrigation canals, and in some instances by desalination of brackish water or use of crops that can utilize saline water. Unused wet soils have been made productive when drained.

On the other hand, soil acidity has been increased (low pH) by the use of commercial fertilizers. Low soil pH has a tendency to make phosphorus and potassium less available to plants as insoluble iron and aluminum phosphates and creates a nutritional deficiency of calcium and magnesium. It is in itself toxic to some plants. This problem often is correctable by the addition of lime to the soil. Irrigation has adversely affected some soils to which the water was applied. Both the use of water by plants and evaporation of water to the atmosphere has caused the increase of salts in many irrigated soils. Many drained lands have lost their utility as nesting grounds to migratory water fowl.

Some soils have been covered by man's wastes in one way or another. Wasteland, a result of strip mine operations and mine waste dumps, often indicated as "made land" on soil survey maps, covers sizeable acreages.

Bethlehem Steel at Lebanon, Pa., through the efforts of its chief forester, has been successful in modifying land of this type. In this case the tailings created an area of devastation in the vicinity of the mine. During dry weather cracks developed in its surface and during wet weather water stood on it for long periods of time. It was seeded and within three years considerable change had occurred (Figure 6.12). Vegetation developed rapidly and the land became productive. There are many areas where similar reclamation projects of land have taken place. There are many more that have not been so cared for and improved (strip mine areas, for example).

Figure 6.12. A mine waste area, dusty when dry and an area of stand-ing water when wet, was planted to trees and shrubs by the Bethlehem Steel Corporation foresters. The results are illustrated in the sequence of three photos which cover a span of three years. "Made land" need not be an unproductive eyesore.

A good comparison of virgin with cultivated soils in New England as to changes made by man's use is shown in Table 6.4. In almost every instance improvement in the mineral content is evident; how-ever, in every case the organic matter content has decreased.

Pesticides and the Soil

Pesticides in agricultural use have had a phenomenal growth. In the years following the discovery of the insecticidal properties of DDT, its benefits were spread worldwide. It has been of incalculable value to man. Yet in recent years it has been decried as the most insidious of the many chemicals contaminating the environment. As one of the persistent pesticides it becomes concentrated in the birds and animals

TABLE 6.4

Average Nutrient Content of Selected New England "Potato" soils under cultivation and in virgin state.

Kinds of soils	Fields sampled	Readily soluble P_2O_5	Exchangeable cations			Organic matter	Acidity
			K_2O	CaO	MgO		
	Number	Pounds per acre	Pounds per acre	Pounds per acre	Pounds per acre	Percent	pH
Maine							
Cultivated	160	217	407	2,095	186	4.91	4.92
Virgin	10	34	68	1,136	137	-	-
New Hampshire							
Cultivated	15	224	-	1,019	50	4.12	4.28
Virgin	15	47	-	1,089	58	5.51	4.56
Rhode Island							
Cultivated	16	479	662	2,786	49	4.24	4.89
Virgin	10	39	111	807	50	5.85	4.80

at the top of the food chain. It has been found in milk and subsequently its use has been restricted in many milksheds. Its complete ban has been proposed by the Wisconsin Natural Resources Department. The U.S. Department of Agriculture has reduced its recommendations as to quantities of DDT to be used and in some cases it has been recommended only for very specific uses such as for the protection of wooden structures in or in contact with the soil.

The United States production in recent years has been in the vicinity of 270 million pounds. The use of DDT has declined markedly to the point where only about half that amount is now being used in the United States. Then why the concern, since its use is rapidly decreasing? Many of these chlorinated hydrocarbons of which DDT is but one, are sufficiently stable to remain in the soil for months and even years, decreasing percentagewise only slightly over the period of time. It is this accumulation of residues in the soil, taken up by plants and passed on to animals and man, that causes the concern. These pesticides, like many of the minerals commonly found in the soil under natural conditions or added as fertilizer elements by man, have become a part of the soil, and in essence have caused a modification of the soil. The concern, however, is not so much that they are added to the soil, but rather because of the fact that the chlorinated hydrocarbons are so slowly degraded in the soil, move from plant to animal and up the food chain to higher animals and to man, where it is stored in body fat and can become a highly toxic substance as body fat is utilized.

Concluding Statement

The soil unquestionably is a resource of prime importance to man. Some would go so far as to say that it is the most important resource on the face of the earth. Undeniably, it is necessary for feed and food production. And equally undeniably it has been abused by those who have allowed soil erosion to progress, and have not replenished the mineral elements taken from the soil. By some it has been coddled and "fed" with mineral and organic fertilizers to the point that its quality is superior to the original. Even complete waste materials not considered as soil at all, have been reclaimed and made productive in one way or another.

Where the soil has proven inadequate for specific agricultural and engineering uses it has been modified to serve the specific use required of it. Breaking up a hardpan where it interfered with root development or making the soil impermeable when irrigation canals required it have been mentioned. Nonetheless there is a limit to these kinds of manipulations. Cement was forced into the unstable soils beneath the Tower of Pisa to make a stable base and hence save the historic tower from toppling. A similar procedure would be out of the question economically for unstable soils of any extent upon which a modern highway has been built. Relocation, after study of a soils report, to a new site on suitable stable soils would be more economical. And better still, had the site been selected with the aid of the soils report before construction had started its unstable characteristics, whether a result of saturation by water or inherent physical makeup of the soils themselves, would have been recognized in time to take appropriate action.

A knowledge of what makes up a soil, how it develops, the extent of its distribution and why, are important points for an understanding of the character, complexity, and variety of our important soils resource. Planning for any use of the soil from agriculture to zoning is best preceded by a thorough knowledge of the soils of the area and this may most readily be gained from the detailed studies of the soil survey reports.

SELECTED REFERENCES

ALBRECHT, WILLIAM A., "Physical, Chemical and Biochemical Changes in the Soil Comunity," *Man's Role in Changing the Face of the Earth,* William L. Thomas, Jr., ed. Chicago: The University of Chicago Press, 1965, pp. 648-673.

Bartelli, L. J., Klinebiel, A. A., Baird, J. V. and Heddleson, M. R., eds. *Soil Surveys and Land Use Planning.* American Society of Agronomy, Madison: 1966.

BRETH, STEVEN, ed. *Pesticides and their Effects on Soils and Water,* American Society Agronomy Special Publication No. 8, Madison: 1966.

CHRISTENSEN, RONDO A. and DEGEORGIO, FRED. "The Green Belt Amendment and Its Probable Impact on Assessed Values, Taxes, and Mill Levies in Salt Lake County." *Utah Science* September 1968, pp. 64-69.

DAHL, A. R., GLENN, J. R., and HANDY, R. L. "Engineering Properties of Missouri River Floodplain Soils," Engineering Research Institute Report, Ames, Ia: n.d.

FRANCIS, JAMES M. "Mine-Waste Reclamation via Vegetative Stabilization." Prepared for presentation to the American Society of Agronomy at its annual convention in Kansas City, Missouri, Nov. 1964.

PAYLORE, PATRICIA. Arid Lands Research—a Selected Bibliography, 1891-1965. Tucson: University of Arizona, 1965. Contains 1609 entries many of which relate to arid land soil.

Thomas, Richard E., Cohen, Jesse M., and Bendixen, Thomas W. *Pesticides in Soil and Water—an Annotated Bibliography.* Engineering Section, Robert A. Taft Sanitary Engineering Center, Cincinnati: 1964. The bibliography has 437 entries.

WADLEIGH, C. H. and DYAL, R. S. "Soils and Pollution," *Agronomy and Health,* R. E. Blaser, ed. American Society of Agonomy Special Publication, No. 16, 1970, pp. 9-19.

Glossary of Soil Science Terms

A

ABC soil—A soil with a distinctly developed profile, including A, B, and C horizons.

AC soil—A soil having a profile containing only A and C horizons with no clearly developed B horizon.

acid soil—A soil with a preponderance of hydrogen ions, and probably of aluminum, in proportion to hydroxyl ions. Specifically, soil with a pH value < 7.0. For most practical purposes a soil with a pH value < 6.6. The pH values obtained vary greatly with the method used and, consequently, there is no unanimous agreement as to what constitutes an acid soil. (The term is usually applied to the surface layer or to the root zone unless specified otherwise.)

adsorption complex—The group of substances in soil capable of adsorbing other materials. Organic and inorganic colloidal substances form the greater part of the adsorption complex; the noncolloidal materials, such as silt and sand, exhibit adsorption but to a much lesser extent than the colloidal materials.

aeration, soil—The process by which air in the soil is replaced by air from the atmosphere. In a well-aerated soil, the soil air is very similar in composition to the atmosphere above the soil. Poorly aerated soils usually contain a much higher percentage of carbon dioxide and a correspondingly lower percentage of oxygen than the atmosphere above the soil. The rate of aeration depends largely on the volume and continuity of pores within the soil.

air porosity—The proportion of the bulk volume of soil that is filled with air at any given time or under a given condition such as a specified moisture tension. Usually the large pores; that is, those drained by a tension of less than approximately 100 cm of water.

alkali soil—(i) A soil with a high degree of alkalinity (pH of 8.5 or higher) or with a high exchangeable sodium content (15% or more of the ex-

Reprinted by permission from *Soil Science Society of America Proceedings*, Vol. 29, No. 3, May-June 1965, pages 330-351, 677 South Segoe Road, Madison, Wisconsin 53711 USA.

change capacity), or both. (ii) A soil that contains sufficient alkali (sodium) to interfere with the growth of most crop plants. See **saline-alkali soil** and **sodic soil.**

alkaline soil—Any soil that has a pH > 7.0. See **reaction, soil.**

Alluvial soil—(i) A soil developing from recently deposited alluvium and exhibiting essentially no horizon development or modification of the recently deposited materials. (ii) When capitalized the term refers to a great soil group of the azonal order consisting of soils with little or no modification of the recent sediment in which they are forming (indicated by absence of a B horizon).

Alpine Meadow soils—A great soil group of the intrazonal order, comprised of dark soils of grassy meadows at altitudes above the timberline.

amendment, soil—(i) An alteration of the properties of a soil, and thereby of the soil, by the addition of substances such as lime, gypsum, sawdust, etc., to the soil for the purpose of making the soil more suitable for the production of plants. (ii) Any such substance used for this purpose. Strictly speaking, fertilizers constitute a special group of soil amendments.

anaerobic—(i) The absence of molecular oxygen. (ii) Growing in the absence of molecular oxygen (such as anaerobic bacteria). (iii) Occurring in the absence of molecular oxygen (as a biochemical process).

anion-exchange capacity—The sum total of exchangeable anions that a soil can adsorb. Expressed as milliequivalents per 100 grams of soil (or of other adsorbing material such as clay).

azonal soils—Soils without distinct genetic horizons. A soil order.

B

base-saturation percentage—The extent to which the adsorption complex of a soil is saturated with exchangeable cations other than hydrogen. It is expressed as a percentage of the total cation-exchange capacity.

BC soil—A soil profile with B and C horizons but with little or no A horizon.

biological interchange—The interchange of elements between organic and inorganic states in a soil or other substrate through the agency of biological activity. It results from biological decomposition of organic compounds and the liberation of inorganic materials (mineralization); and, from the utilization of inorganic materials in the synthesis of microbial tissue (immobilization). Both processes commonly proceed continuously in soils.

biosequence—A sequence of related soils that differ, one from the other, primarily because of differences in kinds and numbers of *soil organisms* as a soil-forming factor.

Black Earth—A term used by some as synonymous with "Chernozem;" by others (in Australia) to describe self-mulching black clays.

Black Soils—A term used in Canada to describe soils with dark-colored surface horizons of the black (Chernozem) zone; includes Black Earth or Chernozem, Wiesenboden, Solonetz, etc.

bleicherde—The light-colored, leached A2 horizon of Podzol soils.

Bog soil—A great soil group of the intrazonal order and hydromorphic suborder. Includes muck and peat.

Brown Earths—Soils with a mull horizon but having no horizon of accumulation of clay or sesquioxides. (Generally used as a synonym for "Brown Forest soils" but sometimes for similar soils acid in reaction.)

Brown Forest soils–A great soil group of the intrazonal order and calcimorphic suborder, formed on calcium-rich parent materials under deciduous forest, and possessing a high base status but lacking a pronounced illuvial horizon. (A much more narrow group than the European Brown Forest or Braunerde.)

Brown Podzolic soils–A zonal great soil group similar to Podzols but lacking the distinct A2 horizon characteristic of the Podzol group. (Some American soil taxonomists prefer to class this soil as a kind of Podzol and not as a distinct great soil group.)

Brown soils–A great soil group of the temperate to cool arid regions, composed of soils with a brown surface and a light-colored transitional subsurface horizon over calcium carbonate accumulation.

Brunizem–Synonymous with **Prairie soils.**

buffer compounds, soil–The clay, organic matter, and compounds such as carbonates and phosphates which enable the soil to resist appreciable change in pH.

buried soil–Soil covered by an alluvial, loessal, or other deposit, usually to a depth greater than the thickness of the solum.

C

calcareous soil–Soil containing sufficient calcium carbonate (often with magnesium carbonate) to effervesce visibly when treated with cold 0.1N hydrochloric acid.

calcification–(Obsolete) The process or processes of soil formation in which the surface soil is kept sufficiently supplied with calcium to saturate the soil colloid, or the process of accumulation of calcium in some horizon of the profile, such as the carbonate horizon of Chernozems.

calciphytes–Plants that require or tolerate considerable amounts of calcium or, are associated with soils rich in calcium.

caliche–(i) A layer near the surface, more or less cemented by secondary carbonates of calcium or magnesium precipitated from the soil solution. It may occur as a soft thin soil horizon, as a hard thick bed just beneath the solum, or as a surface layer exposed by erosion (see **croute calcaire**). Not a geologic deposit. (ii) Alluvium cemented with sodium nitrate, chloride, and/or other soluble salts in the nitrate deposits of Chile and Peru.

capillary fringe–A zone just above the water table (zero gauge pressure) that remains almost saturated. (The extent and the degree of definition of the capillary fringe depends upon the size-distribution of pores.)

capillary water–(Obsolete) The water held in the "capillary" or *small* pores of a soil, usually with a tension > 60 cm of water. See **moisture tension.**

carbon cycle–The sequence of transformations whereby carbon dioxide is fixed in living organisms by photosynthesis or by chemosynthesis, liberated by respiration and by the death and decomposition of the fixing organism, used by heterotrophic species, and ultimately returned to its original state.

carbon-nitrogen ratio–The ratio of the weight of organic carbon to the weight of total nitrogen in a soil or in organic material. It is obtained by dividing the percentage of organic carbon (C) by the percentage of total nitrogen (N).

category—Any one of the ranks of the system of soil classification in which soils are grouped on the basis of their characteristics.

catena—A sequence of soils of about the same age, derived from similar parent material, and occurring under similar climatic conditions, but having different characteristics due to variation in *relief* and in *drainage*. See **clinosequence** and **toposequence.**

cation exchange—The interchange between a cation in solution and another cation on the surface of any surface-active material such as clay colloid or organic colloid.

cation-exchange capacity—The sum total of exchangeable cations that a soil can adsorb. Sometimes called "total-exchange capacity," "base-exchange capacity," or "cation-adsorption capacity." Expressed in milliequivalents per 100 grams of soil (or of other adsorbing material such as clay).

cemented—Indurated; having a hard, brittle consistency because the particles are held together by cementing substances such as humus, calcium carbonate, or the oxides of silicon, iron, and aluminum. The hardness and brittleness persist even when wet.

channery—In Scotland and Ireland, gravel; in the United States, thin, flat fragments of limestone, sandstone, or schist up to 6 inches in major diameter. See **coarse fragments.**

chemically precipitated phosphorus—Relatively insoluble phosphorus compounds resulting from reactions between constituents in solution to form chemically homogeneous particles of the solid phase. Examples are: calcium and magnesium phosphates which are precipitated above a pH of about 6.0 to 6.5 (if calcium and magnesium are present); and, iron and aluminum phosphates which are precipitated below a pH of about 5.8 to 6.1 at which many iron and aluminum compounds are soluble. A form of fixed phosphate. See chemisorbed phosphorus.

chemisorbed phosphorus—Phosphorus adsorbed or precipitated on the surface of clay minerals or other crystalline materials as a result of the attractive forces between the phosphate ion and constituents in the surface of the solid phase.

Chernozem—A zonal great soil group consisting of soils with a thick, nearly black or black, organic matter-rich A horizon high in exchangeable calcium, underlain by a lighter colored transitional horizon above a zone of calcium carbonate accumulation; occurs in a cool subhumid climate under a vegetation of tall and midgrass prairie.

Chestnut soil—A zonal great soil group consisting of soils with a moderately thick, dark-brown A horizon over a lighter colored horizon that is above a zone of calcium carbonate accumulation.

chroma—The relatively purity, strength, or saturation of a color; directly related to the dominance of the determining wavelength of the light and inversely related to grayness; one of the three variables of color. See **Munsell color system, hue,** and **value, color.**

chronosequence—A sequence of related soils that differ, one from the other, in certain properties primarily as a result of *time* as a soil-forming factor.

class, soil—A group of soils having a definite range in a particular property such as acidity, degree of slope, texture, structure, land-use capability, degree of erosion, or drainage. See **soil texture** and **soil structure.**

classification, soil—The systematic arrangement of soils into groups or categories on the basis of their characteristics. Broad groupings are made on the basis of general characteristics and subdivisions on the basis of more detailed differences in specific properties. In the United States the following system has been used for a number of years (*from* Soil Survey Staff, SCS, USDA. 1960. Soil classification: A comprehensive system—7th approximation, p. 9. U.S. Government Printing Office, Washington).

clay—(i) A soil separate consisting of particles $<$ 0.002 mm in equivalent diameter. See **soil separates**. (ii) A textural class. See **soil texture**.

clay mineral—(i) Naturally occuring inorganic crystalline material found in soils and other earthy deposits, the particles being of clay size; that is $<$ 0.002 mm in diameter. (ii) Material as described under (i), but not limited by particle size.

claypan—A dense, compact layer in the subsoil having a much higher clay content than the overlying material, from which it is separated by a sharply defined boundary; formed by downward movement of clay or by synthesis of clay in place during soil formation. Claypans are usually hard when dry, and plastic and sticky when wet. Also, they usually impede the movement of water and air, and the growth of plant roots.

climatic index—A simple, single numerical value that expresses climatic relationships; for example, the numerical value obtained in Transeau's precipitation-evaporation ratio.

climosequence—A sequence of related soils that differ, one from the other, in certain properties primarily as a result of the effect of *climate* as a soil-forming factor.

clinosequence—A group of related soils that differ, one from the other, in certain properties primarily as a result of the effect of the degree of slope on which they were formed. See **toposequence**.

coarse fragments—Rock or mineral particles $>$ 2.0 mm in diameter. The following names are used for coarse fragments in soils.*

Fragments†		Descriptive terms applied to fragments that have:		
Shape	Material	Diameters less than 3 inches	Diameters from 3 to 10 inches	Diameters more than 10 inches
rounded or subrounded	all kinds of rock	gravelly	cobbly	stony‡
irregular and angular	chert	cherty angular gravelly	coarse cherty angular cobbly	stony stony
	other than chert			
		Lengths up to 6 inches	Lengths from 6 to 15 inches	Lengths over 15 inches
thin and flat	limestone, sandstone, or schist	channery	flaggy	stony
	slate	slaty	flaggy	stony
	shale	shaly	flaggy	stony

*From: Soil Survey Staff, SCS, USDA. 1951. Soil survey manual, U.S. Department of Agriculture Handbook 18, p. 214. U.S. Government Printing Office, Washington. †The adjectives describing fragments are also applied to lands and soils when they have significant amounts of such fragments. ‡"bouldery" is sometimes used when stones are larger than 24 inches.

coarse texture—The texture exhibited by sands, loamy sands, and sandy loams except very fine sandy loam. A soil containing large quantities of these textural classes (United States usage).

colloid—Soil, from the Greek, the term "colloid" refers to organic and inorganic matter of very small size and a corresponding high surface area per unit mass.

colluvium—A deposit of rock fragments and soil material accumulated at the base of steep slopes as a result of gravitational action. See **creep**.

color—See **Munsell color system.**

compost—Organic residues, or a mixture of organic residues and soil, that have been piled, moistened, and allowed to undergo biological decomposition. Mineral fertilizers are sometimes added. Often called "artificial manure" or "synthetic manure" if produced primarily from plant residues.

concretion—A local concentration of a chemical compound, such as calcium carbonate or iron oxide, in the form of a grain or nodule of varying size, shape, hardness, and color.

consistency—(i) The resistance of a material to deformation or rupture. (ii) The degree of cohesion or adhesion of the soil mass. Terms used for describing consistency at various soil moisture contents are:
WET SOIL—nonsticky, slightly sticky, sticky, very sticky, nonplastic, slightly plastic, plastic, and very plastic.
MOIST SOIL—loose, very friable, friable, firm, very firm, and extremely firm.
DRY SOIL—loose, soft, slightly hard, hard, very hard, and extremely hard.
CEMETATION—weakly cemented, strongly cemented, and indurated.

coppice mound—A small mound of stabilized soil material around desert shrubs. (A microrelief term.)

cradle knoll—A small knoll formed by earth that is raised and left by an uprooted tree. (A microrelief term.)

creep—Slow mass movement of soil and soil material down relatively steep slopes primarily under the influence of gravity, but facilitated by saturation with water and by alternate freezing and thawing.

crotovina—A former animal burrow in one soil horizon that has been filled with organic matter or material from another horizon (also spelled "krotovina").

crumb structure—A structural condition in which most of the peds are crumbs. See **soil structure types.**

crust—A surface layer on soils, ranging in thickness from a few millimeters to perhaps as much as an inch, that is much more compact, hard, and brittle, when dry, than the material immediately beneath it.

cultivation—A tillage operation used in preparing land for seeding or transplanting or later for weed control and for loosening the soil.

cyclic salt—Salt deposited on the soil by wind blowing off the sea or off inland salt lakes.

D

D layer—(Obsolete) An unconsolidated geological stratum, beneath the solum of some soils, that does not conform to the parent material from which the overlying soil was formed even though it may influence the genesis or behavior of the soil. See **soil horizon.**

Dark Gray Gleysolic soil—A term used in Canada to describe an intrazonal group of imperfectly to poorly-drained forested soils having dark-gray

A horizons, moderately high in organic matter, underlain by mottled gray or brownish gleyed mineral horizons. They have a low degree of textural differentiation.

decalcification—The removal of calcium carbonate or calcium ions from the soil by leaching.

deflocculate—(i) To separate the individual components of compound particles by chemical and/or physical means. (ii) To cause the particles of the *disperse phase* of a colloidal system to become suspended in the *dispersion medium*.

Degraded Chernozem—A zonal great soil group consisting of soils with a very dark brown or black A1 horizon underlain by a dark gray, weakly expressed A2 horizon and a brown B (?) horizon; formed in the forest-prairie transition of cool climates.

denitrification—The biochemical reduction of nitrate or nitrate to gaseous nitrogen either as molecular nitrogen or as an oxide of nitrogen.

Depression Podzol—Poorly drained depressional soils of the grassland and parkland regions of Canada with bleached A2 horizons and finer-textured B horizons. Also referred to as **Bluff Podzol, Meadow Podzol,** or **Slough Podzol.**

desert crust—A hard layer, containing calcium carbonate, gypsum, or other binding material, exposed at the surface in desert regions.

desert pavement—The layer of gravel or stones left on the land surface in desert regions after the removal of the fine material by wind erosion.

Desert soil—A zonal great soil group consisting of soils with a very thin, light-colored surface horizon, which may be vesicular and is ordinarily underlain by calcareous material; formed in arid regions under sparse shrub vegetation.

desert varnish—A glossy sheen or coating on stones and gravel in arid regions.

diatoms—Algae having siliceous cell walls that persist as a skeleton after death. Any of the microscopic unicellular or colonial algae constituting the class Bacillariaceae. They occur abundantly in fresh and salt waters and their remains are widely distributed in soils.

diatomaceous earth—A geologic deposit of fine, grayish siliceous material composed chiefly or wholly of the remains of diatoms. It may occur as a powder or as a porous, rigid material.

disperse—(i) To break up compound particles, such as aggregates, into the individual component particles. (ii) To distribute or suspend fine particles, such as clay, in or throughout a dispersion medium, such as water.

duff mull—A type of forest humus transitional between mull and mor; H and F layers as well as the A1 horizon.

dust mulch—A loose, finely granular, or powdery condition on the surface of the soil, usually produced by shallow cultivation.

dy—Finely divided, partially decomposed organic material accumulated in peat soils in the transition zone between the peat and the underlying mineral material.

E

ectodynamomorphic soils—Soils with properties that have been produced or influenced mainly by factors other than parent material. See **endodynamomorphic soils.**

edaphic—(i) Of or pertaining to the soil. (ii) Resulting from or influenced by factors inherent in the soil or other substrate, rather than by climatic factors.

edaphology—The science that deals with the influence of soils on living things, particularly plants, including man's use of land for plant growth.

eluvial horizon—A soil horizon that has been formed by the process of eluviation. See **illuvial horizon.**

eluviation—The removal of soil material in suspension (or in solution) from a layer or layers of a soil. (Usually, the loss of material in *solution* is described by the term "leaching." See **illuviation** and **leaching.**

endodynamomorphic soils—Soils with properties that have been influenced primarily by parent material.

erosion—(i) The wearing away of the land surface by running water, wind, ice, or other geological agents, including such processes as gravitational creep. (ii) Detachment and movement of soil or rock by water, wind, ice, or gravity. The following terms are used to describe different types of water erosion:

accelerated erosion—Erosion much more rapid than normal, natural, geological erosion, primarily as a result of the influence of the activities of man or, in some cases, of animals.

geological erosion—The normal or natural erosion caused by geological processes acting over long geologic periods and resulting in the wearing away of mountains, the building up of flood plains, coastal plains, etc. Synonymous with *natural erosion.*

gully erosion—The erosion process whereby water accumulates in narrow channels and, over short periods, removes the soil from this narrow area to considerable depths, ranging from 1 or 2 feet to as much as 75 to 100 feet.

natural erosion—Wearing away of the earth's surface by water, ice, or other natural agents under natural environmental conditions of climate, vegetation, etc., undisturbed by man. Synonymous with *geological erosion.*

normal erosion—The gradual erosion of land used by man which does not greatly exceed natural erosion. See *natural erosion.*

rill erosion—An erosion process in which numerous small channels of only several inches in depth are formed; occurs mainly on recently cultivated soils. See **rill.**

sheet erosion—The removal of a fairly uniform layer of soil from the land surface by runoff water.

splash erosion—The spattering of small soil particles caused by the impact of raindrops on very wet soils. The loosened and spattered particles may or may not be subsequently removed by surface runoff.

erosion classes—A grouping of erosion conditions based on the degree of erosion or on characteristic patterns. (Applied to accelerated erosion; not to normal, natural, or geological erosion.) Four erosion classes are recognized for water erosion and three for wind erosion. Specific definitions for each vary somewhat from one climatic zone, or major soil group, to another. (For details see Soil Survey Staff, SCS, USDA. 1951. Soil survey manual, U.S. Dept. Agriculture Handbook 18. U.S. Government Printing Office, Washington.)

exchange capacity—The total ionic charge of the adsorption complex active in the adsorption of ions. See **anion-exchange capacity** and **cation-exchange capacity.**

exchangeable phosphate—The phosphate anion reversibly attached to the surface of the solid phase of the soil in such form that it may go into solution by anionic equilibrium reactions with isotopes of phosphorus or with other anions of the liquid phase without solution of the colloid phase to which it was attached.

exchangeable potassium—The potassium that is held by the adsorption complex of the soil and is easily exchanged with the cation of neutral non-potassium salt solutions.

F

F layer—A layer of partially decomposed litter with portions of plant structures still recognizable. Occurs below the L layer (O11 horizon) on the forest floor in forest soils. It is the fermentation layer or the O12 layer. See **L layer** and **soil horizon.**

family, soil—In soil classification one of the categories intermediate between the great soil group and the soil series. See **classification, soil.**

field capacity (field moisture capacity)—(Obsolete in technical work.) The percentage of water remaining in a soil 2 or 3 days after having been saturated and after free drainage has practically ceased. (The percentage may be expressed on the basis of weight or volume.)

film water—A layer of water surrounding soil particles and varying in thickness from 1 or 2 to perhaps 100 or more molecular layers. Usually considered as that water remaining after drainage has occurred, because it is not distinguishable in saturated soils.

fine texture—Consisting of or containing large quantities of the fine fractions, particularly of silt and clay. (Includes all clay loams and clays; that is, clay loam, sandy clay loam, silty clay loam, sandy clay, silty clay, and clay textural classes. Sometimes subdivided into clayey texture and moderately fine texture.) See **soil texture.**

firm—A term describing the consistency of a moist soil that offers distinctly noticeable resistance to crushing but can be crushed with moderate pressure between the thumb and forefinger. See **consistency.**

fixed phosphorus—(i) That phosphorus which has been changed to a less soluble form as a result of reaction with the soil; moderately available phosphorus. More specifically, that quantity of soluble phosphorus compounds which, when added to soil, becomes chemically or biologically attached to the solid phase of soil so as not to be recovered by extracting the soil with a specified extractant under specified conditions. Such extractants include: water, carbonated water, or dilute solutions of strong mineral acids with or without fluoride or other exchangeable anion. (ii) Applied phosphorus that is not absorbed by plants during the first cropping year. (iii) Soluble phosphorus that has become attached to the solid phase of the soil in forms highly unavailable to crops; unavailable phosphorus; phosphorus in other than readily or moderately available forms.

flow velocity (of water in soil)—The volume of water transferred per unit of time and per unit of area normal to the direction of the net flow.

forest soils—(i) Soils developed under forest vegetation. (ii) Soils formed in temperate climates under forest vegetation (European usage).

fragipan—A natural subsurface horizon with high bulk density relative to the solum above, seemingly cemented when dry, but when moist showing a moderate to weak brittleness. The layer is low in organic matter, mottled, slowly or very slowly permeable to water, and usually shows occasional or frequent bleached cracks forming polygons. It may be found in profiles of either cultivated or virgin soils but not in calcareous material.

friable—A consistency term pertaining to the ease of crumbling of soils. See **consistency**.

G

genetic—Resulting from, or produced by, soil-forming processes; for example, a genetic soil profile or a genetic horizon.

gilgai—The microrelief of soils produced by expansion and contraction with changes in moisture. Found in soils that contain large amounts of clay which swells and shrinks considerably with wetting and drying. Usually a succession of microbasins and microknolls in nearly level areas or of microvalleys and microridges parallel to the direction of the slope.

glacial soil—(Obsolete) A soil derived from glacial drift.

Gleyzation—A soil-forming process resulting in the development of gley soils. See **Humic Gley soil** and **Dark Gray Gleysolic soil.**

Gley soil—(Obsolete in the USA) Soil developed under conditions of poor drainage resulting in reduction of iron and other elements and in gray colors and mottles.

granule—A natural soil aggregate or ped which is relatively nonporous. See **soil structure.**

gravitational water—Water which moves into, through, or out of the soil under the influence of gravity.

Gray-Brown Podzolic soil—A zonal great soil group consisting of soils with a thin, moderately dark A1 horizon and with a grayish-brown A2 horizon underlain by a B horizon containing a high percentage of bases and an appreciable quantity of illuviated silicate clay; formed on relatively young land surfaces, mostly glacial deposits, from material relatively rich in calcium, under deciduous forests in humid temperate regions.

Gray Desert soil—A term used in Russia, and frequently in the United States, synonymously with Desert soil. See **Desert soil.**

Great soil group—One of the categories in the system of soil classification that has been used in the United States for many years. See **classification, soil.**

groundwater—That portion of the total precipitation which at any particular time is either passing through or standing in the soil and the underlying strata and is free to move under the influence of gravity. See **water table.**

Ground-Water Laterite soil—A great soil group of the intrazonal order and hydromorphic suborder, consisting of soils characterized by hardpans or concretional horizons rich in iron and aluminum (and sometimes manganese) that have formed immediately above the water table.

Ground-Water Podzol soil—A great soil group of the intrazonal order and hydromorphic suborder, consisting of soils with an organic mat on the surface over a very thin layer of acid humus material underlain by a whitish-gray leached layer, which may be as much as 2 or 3 feet in thickness, and is underlain by a brown, or very dark-brown, cemented hardpan layer; formed under various types of forest vegetation in cool to tropical, humid climates under conditions of poor drainage.

gyttja—Sedimentary peat consisting mainly of plant and animal residues precipitated from standing water.

H

H layer—A layer occurring in mor humus consisting of well-decomposed organic matter of unrecognizable origin. The O2 horizon. See **soil horizon.**

Half-Bog soil—A great soil group, of the intrazonal order and hydromorphic suborder consisting of soil with dark-brown or black peaty material over grayish and rust mottled mineral soil; formed under conditions of poor drainage under forest, sedge, or grass vegetation in cool to tropical humid climates.

halomorphic soil—A suborder of the intrazonal soil order, consisting of saline and alkali soils formed under imperfect drainage in arid regions and including the great soil groups Solonchak or Saline soils, Solonetz soils, and Soloth soils.

hardpan—A hardened soil layer, in the lower A or in the B horizon, caused by cementation of soil particles with organic matter or with materials such as silica, sesquioxides, or calcium carbonate. The hardness does not change appreciably with changes in moisture content and pieces of the hard layer do not slake in water. See **caliche** and **claypan.**

heavy soil—(Obsolete in scientific use.) A soil with a high content of the fine separates, particularly clay, or one with a high drawbar pull and hence difficult to cultivate. See **fine texture.**

hue—One of the three variables of color. It is caused by light of certain wavelengths and changes with the wavelength. See **Munsell color system, chroma,** and **value, color.**

humic acid—A mixture of variable or indefinite composition of dark-colored organic substances, precipitated upon acidification of a dilute-alkali extract from soil. (Used by some workers to designate only the alcohol-soluble portion of this precipitate.) (In chemical literature, it is sometimes used to designate a preparation obtained by the treatment of sugars with mineral acids.)

Humic Gley soil—Soil of the intrazonal order and hydromorphic suborder that includes Wisenboden and related soils, such as Half-Bog soils, which have a thin muck or peat O2 horizon and an A1 horizon. Developed in wet meadow and in forested swamps.

humification—The processes involved in the decomposition of organic matter and leading to the formation of humus.

humin—The fraction of the soil organic matter that is not dissolved upon extraction of the soil with dilute alkali.

humus—(i) That more or less stable fraction of the soil organic matter remaining after the major portion of added plant and animal residues have decomposed. Usually it is dark colored. (ii) Includes the F and H layers in undisturbed forest soils. See **soil organic matter, mor, mull,** and **soil horizon.**

hydrogenic soil—Soil developed under the influence of water standing within the profile for considerable periods; formed mainly in cold, humid regions.

hydromorphic soils—A suborder of intrazonal soils, consisting of seven great soil groups, all formed under conditions of poor drainage in marshes, swamps, seepage areas, or flats. See **classification, soil.**

hygroscopic water—Water adsorbed by a dry soil from an atmosphere of high relative humidity, water remaining in the soil after "air-drying," or water held by the soil when it is in equilibrium with an atmosphere of a specified relative humidity at a specified temperature, usually 98% relative humidity at 25C.

I

illite—A hydrous mica. See **hydrous mica.**

illuvial horizon—A soil layer or horizon in which material carried from an overlying layer has been precipitated from solution or deposited from suspension. The layer of accumulation. See **eluvial horizon.**

illuviation—The process of deposition of soil material removed from one horizon to another in the soil; usually from an upper to a lower horizon in the soil profile. See **eluviation.**

immature soil—A soil with indistinct or only slightly developed horizons because of the relatively short time it has been subjected to the various soil-forming processes. A soil that has not reached equilibrium with its environment.

infiltration rate—A soil characteristic determining or describing the maximum rate at which water can enter the soil under specified conditions, including the presence of an excess of water. It has the dimensions of velocity (i.e., cm^3 cm^{-2} sec^{-1} = cm sec^{-1}) (Formerly, the infiltration capacity.) See **infiltration velocity.**

infiltration velocity—The *actual* rate at which water is entering the soil at any given time. It may be less than the maximum (the **infiltration rate**) because of a limited supply of water (rainfall or irrigation). It has the same units as the ifiltration rate. See **infiltration rate.**

intergrade—A soil that possesses moderately well-developed distinguishing characteristics of two or more genetically related great soil groups.

intrazonal soils—(i) One of the three orders in soil classification. See **classification, soil.** (ii) A soil with more or less well-developed soil characteristics that reflect the dominating influence of some local factor of relief, parent material, or age, over the normal effect of climate and vegetation.

intrinsic permeability—The property of a porous material that relates to the ease with which gases or liquids can pass through it.

ion activity—The effective concentration of a particular ion in a solution or soil-water system. It is expressed analogously to pH, as "pCa," "pNa," etc.

iron-pan—An indurated soil horizon in which iron oxide is the principal cementing agent.

isomorphous substitution—The replacement of one atom by another of similar size in a crystal lattice without disrupting or changing the crystal structure of the mineral.

K

kaolin—(i) An aluminosilicate mineral of the 1:1 crystal lattice group; that is, consisting of one silicon tetrahedral layer and one aluminum oxide-hydroxide octahedral layer. (ii) The 1:1 group or family of aluminosilicates.

L

L layer (litter)—The surface layer of the forest floor consisting of freshly fallen leaves, needles, twigs, stems, bark, and fruits. This layer may be very thin or absent during the growing season. The O1 horizon. See **soil horizon.**

lacustrine soil—(Obsolete) Soil formed on or from lacustrine deposits.

land classification—The arrangement of land units into various categories based upon the properties of the land or its suitability for some particular purpose.

Lateritic soil—A suborder of zonal soils formed in warm, temperate, and tropical regions and including the following great soil groups: Yellow Podzolic, Red Podzolic, Yellowish-Brown Lateritic, and Lateritic. See **classification, soil** and **Latosol.**

Latosol—A suborder of zonal soils including soils formed under forested, tropical, humid conditions and characterized by low silica-sesquioxide ratios of the clay fractions, low base-exchange capacity, low activity of the clay, low content of most primary minerals, low content of soluble constituents, a high degree of aggregate stability, and usually having a red color. See **classification, soil** and **Lateritic soil.**

Leached Saline soils—(i) Soils from which the soluble salts have been removed by leaching. (ii) Soils that have been saline and still possess the major physical characteristics of saline soils but from which the soluble salts have been leached, generally for reclamation.

leaching—The removal of materials in solution from the soil. See **eluviation.**

light soil—(Obsolete in scientific use) A coarse-textured soil; a soil with a low drawbar pull and hence easy to cultivate. See **coarse texture** and **soil texture.**

lime-pan—A hardened layer cemented by calcium carbonate.

lithosequence—A group of related soils that differ, one from the other, in certain properties primarily as a result of differences in the *parent rock* as a soil-forming factor.

Lithosols—A great soil group of azonal soils characterized by an incomplete solum or no clearly expressed soil morphology and consisting of freshly and imperfectly weathered rock or rock fragments.

loamy—Intermediate in texture and properties between fine-textured and coarse-textured soils. Includes all textural classes with the words "loam" or "loamy" as a part of the class name, such as clay loam or loamy sand. See **soil texture.**

loess—Material transported and deposited by wind and consisting of predominantly silt-sized particles.

M

made land—Areas filled with earth, or with earth and trash mixed, usually by or under the control of man. **A miscellaneous land type.**

marl—Soft and unconsolidated calcium carbonate, usually mixed with varying amounts of clay or other impurities.

mature soil—A soil with well-developed soil horizons produced by the natural processes of soil formation and essentially in equilibrium with its present environment.

maximum water-holding capacity—The average moisture content of a disturbed sample of soil, 1 cm high, which it at equilibrium with a water table at its lower surface.

mechanical analysis—(Obsolete) See **particle-size analysis** and **particle-size distribution.**

medium-texture—Intermediate between fine-textured and coarse-textured (soils). (It includes the following textural classes: very fine sandy loam, loam, silt loam, and silt.)

mineral, soil—See **soil mineral.**

mineralization—The conversion of an element from an organic form to an inorganic state as a result of microbial decomposition.

mineralogical analysis—The estimation or determination of the kinds or amounts of minerals present in a rock or in a soil.

mineral soil—A soil consisting predominantly of, and having its properties determined predominantly by, mineral matter. Usually contains $< 20\%$ organic matter, but may contain an organic surface layer up to 30 cm thick.

miscellaneous land type—A mapping unit for areas of land that have little or no natural soil, that are too nearly inaccessible for orderly examination, or that for any reason it is not feasible to classify the soil.—**badland, made land, meander land, mine dumps, mine wash, oil wasteland, river wash, rough broken land, rubble land, scoria land, slickens, stony land, swamp, tidal flats, urban land, volcanic-ash land, and waste land.**

moderately-coarse texture—Consisting predominantly of coarse particles. (In soil textural classification, it includes all the sandy loams except the very fine sandy loam.) See **coarse texture.**

moderately-fine texture—Consisting predominantly of intermediate-size (soil) particles or with relatively small amounts of fine or coarse particles. (In soil textural classification, it includes clay loam, sandy clay loam, and silty clay loam.) See **fine texture.**

montmorillonite—An aluminosilicate clay mineral with a 2:1 expanding crystal lattice; that is, with two silicon tetrahedral layers enclosing an aluminum octahedral layer. Considerable expansion may be caused along the C axis by water moving between silica layers of contiguous units. See **montmorillonite group.**

montmorillonite group—Clay minerals with 2:1 crystal lattice structure; that is, two silicon tetrahedral layers enclosing an aluminum octahedral layer. Consists of montmorillonite, beidellite, nontronite, saponite, and others.

mor—A type of forest humus in which the H layer is present and in which there is practically no mixing of surface organic matter with mineral soil; that is, the transition from the H layer to the A1 horizon is abrupt. (Sometimes differentiated into thick mor, thin mor, granular mor, greasy mor, or felty mor.)

mottled zone—A layer that is marked with spots or blotches of different color or shades of color. The pattern of mottling and the size, abundance, and color contrast of the mottles may vary considerably and should be specified in soil description.

mottling–Spots or blotches of different color or shades of color interspersed with the dominant color.

mountain soils–(Obsolete) Skeletal soils formed mainly by physical weathering in cool, mountain regions.

muck–Highly decomposed organic material in which the original plant parts are not recognizable. Contains more mineral matter and is usually darker in color than peat. See **muck soil, peat,** and **peat soil.**

muck soil–(i) A soil containing between 20 and 50% of organic matter. (ii) An organic soil in which the organic matter is well decomposed (USA usage).

mull–A type of forest humus in which the *F layer* may or may not be present and in which there is no *H layer*. The A1 horizon consists of an intimate mixture of organic matter and mineral soil with gradual transition between the A1 and the horizon beneath. (Sometimes differentiated into firm mull, sand mull, coarse mull, medium mull, and fine mull.)

Munsell color system–A color designation system that specifies the relative degrees of the three simple variables of color: hue, value, and chroma. For example: 10 YR 6/4 is a color (of soil) with a hue = 10 YR, value = 6, and chroma = 4. These notations can be translated into several different systems of color names as desired. See **chroma, hue,** and **value, color.**

N

neutral soil–A soil in which the surface layer, at least to normal plow depth, is neither acid nor alkaline in reaction. See **acid soil, alkaline soil, pH,** and **reaction, soil**

nitrogen assimilation–The incorporation of nitrogen into organic cell substances by living organisms.

nitrogen cycle–The sequence of biochemical changes undergone by nitrogen wherein it is used by a living organism, liberated upon the death and decomposition of the organism, and converted to its original state of oxidation.

nitrogen fixation–The conversion of elemental nitrogen (N_2) to organic combinations or to forms readily utilizable in biological processes.

O

O horizon–See **soil horizon.**

organic soil–A soil which contains a high percentage ($>$ 15 or 20%) of organic matter throughout the solum.

ortstein–An indurated layer in the B horizon of Podzols in which the cementing material consists of illuviated sesquioxides (mostly iron) and organic matter.

P

pans–Horizons or layers, in soils, that are strongly compacted, indurated, or very high in clay content. See **caliche, claypan, fragipan,** and **hardpan.**

pan, genetic–A natural subsurface soil layer of low or very low permeability, with a high concentration of small particles, and differing in certain physical and chemical properties from the soil immediately above or below the pan. See **claypan, fragipan,** and **hardpan,** all of which are genetic pans.

pan, pressure or induced–A subsurface horizon or soil layer having a higher bulk density and a lower total porosity than the soil directly above or below it, as a result of pressure that has been applied by normal tillage operations or by other artificial means. Frequently referred to as plowpan, plowsole, or traffic pan.

parent material—The unconsolidated and more or less chemically weathered mineral or organic matter from which the solum of soils is developed by pedogenic processes.

particle size—The effective diameter of a particle measured by sedimentation, sieving, or micrometric methods.

particle-size analysis—Determination of the various amounts of the different separates in a soil sample, usually by sedimentation, sieving, micrometry, or combinations of these methods.

particle-size distribution—The amounts of the various soil separates in a soil sample, usually expressed as weight percentages.

peat—Unconsolidated soil material consisting largely of undecomposed, or only slightly decomposed, organic matter accumulated under conditions of excessive moisture.

peat soil—An organic soil containing more than 50% organic matter. Used in the United States to refer to the stage of decomposition of the organic matter, "peat" referring to the slightly decomposed or undecomposed deposits and "muck" to the highly decomposed materials. See **peat, muck,** and **muck soil.**

ped—A unit of soil structure such as an aggregate, crumb, prism, block, or granule, formed by natural processes (in contrast with a clod, which is formed artificially).

pedalfer—(Obsolete) A subdivision of a soil order comprising a large group of soils in which sesquioxides increased relative to silica during soil formation.

pedocal—(Obsolete) A subdivision of a soil order comprising a large group of soils in which calcium accumulated during soil formation.

penetrability—The ease with which a probe can be pushed into the soil. (May be expressed in units of distance, speed, force, or work depending on the type of penetrometer used.)

percolation, soil water—The downward movement of water through soil. Especially, the downward flow of water in saturated or nearly saturated soil at hydraulic gradients of the order of 1.0 or less.

permafrost—(i) Permanently frozen material underlying the solum. (ii) A perennially frozen soil horizon.

permafrost table—The upper boundary of the permafrost [See permafrost, (i)], coincident with the lower limit of seasonal thaw.

permeability, soil—(i) The ease with which gases, liquids, or plant roots penetrate or pass through a bulk mass of soil or a layer of soil. Since different soil horizons vary in permeability, the particular horizon under question should be designated. (ii) The property of a porous medium itself that relates to the ease with which gases, liquids, or other substances can pass through it.

pH, hydrolytic—The arithmetical difference between the pH value of a soil as measured on the soil paste and the value obtained on a 1:10 soil-water suspension.

pH, isohydric—The pH value of a soil identical with that of a buffer solution that remains unchanged when mixed with the soil.

pH, soil—The negative logarithm of the hydrogen-ion activity of a soil. The degree of acidity (or alkalinity) of a soil as determined by means of a glass, quinhydrone, or other suitable electrode or indicator at a specified moisture content or soil-water ratio, and expressed in terms of the pH scale.

phase, soil—A subdivision of a soil type or other unit of classification having characteristics that affect the use and management of the soil but which do not vary sufficiently to differentiate it as a separate type. A variation in a property or characteristic such as degree of slope, degree of erosion, content of stones, etc.

physical properties (of soils)—Those characteristics, processes, or reactions of a soil which are caused by physical forces and which can be described by, or expressed in, physical terms or equations. Sometimes confused with and difficult to separate from chemical properties; hence, the terms "physical-chemical" or "physicochemical." Examples of physical properties are bulk density, water-holding capacity, hydraulic conductivity, porosity, pore-size distribution, etc.

phytogenic soils—(Obsolete) Soils developed under the dominant influence of the natural vegetation, mainly in temperate regions.

phytomorphic soils—(Canada) Well-drained soils of an association which have developed under the dominant influence of the natural vegetation characteristic of a region. The zonal soils of an area.

Planosol—A great soil group of the intrazonal order and hydromorphic suborder consisting of soils with eluviated surface horizons underlain by B horizons more strongly eluviated, cemented, or compacted than associated normal soil.

plastic soil—A soil capable of being molded or deformed continuously and permanently, by relatively moderate pressure, into various shapes. See **consistency.**

Podzol—A great soil group of the zonal order consisting of soils formed in cool-temperate to temperate, humid climates, under coniferous or mixed coniferous and deciduous forest, and characterized particularly by a highly-leached, whitish-gray (Podzol) A2 horizon.

podzolization—A process of soil formation resulting in the genesis of Podzols and Podzolic soils.

pore space—Total space not occupied by soil particles in a bulk volume of soil.

porosity—The volume percentage of the total bulk not occupied by solid particles.

potassium fixation—The process of converting exchangeable or water-soluble potassium to moderately soluble potassium; i.e., to a form not easily exchanged from the adsorption complex with a cation of a neutral salt solution.

Prairie soils—A zonal great soil group consisting of soils formed under temperate to cool temperate, humid regions under tall grass vegetation. See **classification, soil.**

profile, soil—A vertical section of the soil through all its horizons and extending into the parent material.

R

R horizon—See **soil horizon.**

reaction, soil—The degree of acidity or alkalinity of a soil, usually expressed as a pH value. Descriptive terms commonly associated with certain ranges in pH are: *extremely acid,* < 4.5; *very strongly acid,* 4.5–5.0; *strongly acid,* 5.1–5.5; *moderately acid,* 5.6–6.0; *slightly acid,* 6.1–6.5; *neutral,* 6.6–7.3; *slightly alkaline,* 7.4–7.8; *moderately alkaline,* 7.9–8.4; *strongly alkaline,* 8.5–9.0; and *very strongly alkaline,* > 9.1.

Red Desert soil—A zonal great soil group consisting of soils formed under warm-temperate to hot, dry regions under desert-type vegetation, mostly shrubs.

red earth—Highly leached, red clayey soils of the humid tropics, usually with very deep profiles that are low in silica and high in sesquioxides.

Red-Yellow Podzolic soils—A combination of the zonal great soil groups, Red Podzolic and Yellow Podzolic, consisting of soils formed under warm-temperate to tropical, humid climates, under deciduous or coniferous forest vegetation and usually, except for a few members of the Yellow Podzolic Group, under conditions of good drainage.

regolith—The unconsolidated mantle of weathered rock and soil material on the earth's surface; loose earth materials above solid rock. (Approximately equivalent to the term "soil" as used by many engineers.)

Regosol—Any soil of the azonal order without definite genetic horizons and developing from or on deep, unconsolidated, soft mineral deposits such as sands, loess, or glacial drift.

Regur—An intrazonal group of dark calcareous soils high in clay, which is mainly montmorillonitic, and formed mainly from rocks low in quartz; occurring extensively on the Deccan Plateau of India.

Rendzina—A great soil group of the intrazonal order and calcimorphic suborder consisting of soils with brown or black friable surface horizons underlain by light gray to pale yellow calcareous material; developed from soft, highly calcareous parent material under grass vegetation or mixed grasses and forest in humid and semiarid climates.

residual soil—(Obsolete) A soil formed from, or resting on, consolidated rock of the same kind as that from which it was formed, and in the same location.

reticulate mottling—A network of streaks of different color; most commonly found in the deeper profiles of Lateritic soils.

S

saline-alkali soil—(i) A soil containing sufficient exchangeable sodium to interfere with the growth of most crop plants and containing appreciable quantities of soluble salts. The exchangeable-sodium percentage is > 15, the conductivity of the saturation extract > 4 millimhos per centimeter (at 25C), and the pH is usually 8.5 or less in the saturated soil. (ii) A saline-alkali soil has a combination of harmful quantities of salts and either a high alkalinity or high content of exchangeable sodium, or both, so distributed in the profile that the growth of most crop plants is reduced. (Often called saline-sodic soil.)

saline soil—A nonalkali soil containing sufficient soluble salts to impair its productivity. (This name was formerly applied to any soil containing sufficient soluble salts to interfere with plant growth.)

salinization—The process of accumulation of salts in soil.

secondary mineral—A mineral resulting from the decomposition of a primary mineral or from the reprecipitation of the products of decomposition of a primary mineral.

Sierozem—A zonal great soil group consisting of soils with pale grayish A horizons grading into calcareous material at a depth of 1 foot or less, and formed in temperate to cool, arid climates under a vegetation of desert plants, short grass, and scattered brush.

silica-sesquioxide ratio—The molecules of silicon dioxide (SiO_2) per molecule of aluminum oxide (Al_2O_3) plus ferric oxide (Fe_2O_3) in clay minerals or in soils.

silt—(i) A soil separate consisting of particles between 0.05 and 0.002 mm in equivalent diameter. See **soil separates**. (ii) A soil textural class. See **soil texture**.

single-grain structure—(Obsolete) A soil structure classification in which the soil particles occur almost completely as individual or primary particles with essentially no secondary particles or aggregates being present. (Usually found only in extremely coarse-textured soils.)

slick spots—Small areas in a field that are slick when wet, due to a high content of alkali or of exchangeable sodium.

sodic soil—(i) A soil that contains sufficient sodium to interfere with the growth of most crop plants. (ii) A soil in which the exchangeable-sodium percentage is 15 or more.

soil—(i) The unconsolidated mineral material on the immediate surface of the earth that serves as a natural medium for the growth of land plants. (ii) The unconsolidated mineral matter on the surface of the earth that has been subjected to and influenced by genetic and environmental factors of: *parent material, climate* (including moisture and temperature effects), *macro-* and *microorganisms,* and *topography,* all acting over a period of *time* and producing a product—soil—that differs from the material from which it is derived in many physical, chemical, biological and morphological properties, and characteristics.

soil air—The soil atmosphere; the gaseous phase of the soil, being that volume not occupied by solid or liquid.

soil alkalinity—The degree or intensity of alkalinity of a soil, expressed by a value > 7.0 on the pH scale.

soil associate—(Obsolete) A term used in Canada to define an individual taxonomic unit of a soil association, particularly of an association of soils derived from similar parent materials; corresponds to a series within a catena.

soil association—(i) A group of defined and named taxonomic soil units occurring together in an individual and characteristic pattern over a geographic region, comparable to plant associations in many ways. (Sometimes called "natural land type.") (ii) A mapping unit used on general soil maps, in which two or more defined taxonomic units occurring together in a characteristic pattern are combined because the scale of the map or the purpose for which it is being made does not require delineation of the individual soils.

soil auger—A tool for boring into the soil and withdrawing a small sample for field or laboratory observation. Soil augers may be classified into several types as follows: (i) those with worm-type bits, uninclosed;

(ii) those with worm-type bits inclosed in a hollow cylinder; and (iii) those with a hollow cylinder with a cutting edge at the lower end.

soil complex—A mapping unit used in detailed soil surveys where two or more defined taxonomic units are so intimately intermixed geographically that it is undesirable or impractical, because of the scale being used, to separate them. A more intimate mixing of smaller areas of individual taxonomic units than that described under soil association.

soil-formation factors—The variable, usually interrelated natural agencies that are active in and responsible for the formation of soil. The factors are usually grouped into five major categories as follows: parent rock, climate, organisms, topography, and time.

soil genesis—(i) The mode of origin of the soil with special reference to the processes or soil-forming factors responsible for the development of the solum, or true soil, from the unconsolidated parent material. (ii) A division of soil science concerned with soil genesis (i).

soil geography—A subspecialization of physical geography concerned with the areal distributions of soil types.

soil horizon—A layer of soil or soil material approximately parallel to the land surface and differing from adjacent genetically related layers in physical, chemical, and biological properties or characteristics such as color, structure, texture, consistency, kinds and numbers of organisms present, degree of acidity or alkalinity, etc. The following table lists the designations and properties of the major soil horizons. Very few if any soils have all of these horizons well developed but every soil has some of them.

Horizon designation	Description
O	Organic horizons of mineral soils. Horizons: (i) formed or forming in the upper part of mineral soils above the mineral part; (ii) dominated by fresh or partly decomposed organic material; and (iii) containing > 30% organic matter if the mineral fraction is > 50% clay, or > 20% organic matter if the mineral fraction has no clay. Intermediate clay content requires proportional organic-matter content.
O1	Organic horizons in which essentially the original form of most vegetative matter is visible to the naked eye. The O1 corresponds to the L (litter) and some F (fermentation) layers in forest soils designations, and to the horizon formerly called Aoo.
O2	Organic horizons in which the original form of most plant or animal matter cannot be recognized with the naked eye. The O2 corresponds to the H (humus) and some F (fermentation) layers in forest soils designations, and to the horizon formerly called Ao.
A	Mineral horizons consisting of: (i) horizons of organic-matter accumulation formed or forming at or adjacent to the surface; (ii) horizons that have lost clay, iron, or aluminum with resultant concentration of quartz or other resistant minerals of sand or silt size; or (iii) horizons dominated by (i) or (ii) above but transitional to an underlying B or C.
A1	Mineral horizons, formed or forming at or adjacent to the surface, in which the feature emphasized is an accumulation

of humified organic matter intimately associated with the mineral fraction.

A2   Mineral horizons in which the feature emphasized is loss of clay, iron, or aluminum, with resultant concentration of quartz or other resistant minerals in sand and silt sizes.

A3   A transitional horizon between A and B, and dominated by properties characteristic of an overlying A1 or A2 but having some subordinate properties of an underlying B.

AB   A horizon transitional between A and B, having an upper part dominated by properties of A and a lower part dominated by properties of B, and the two parts cannot be conveniently separated into A3 and B1.

A & B   Horizons that would qualify for A2 except for included parts constituting < 50% of the volume that would qualify as B.

AC   A horizon transitional between A and C, having subordinate properties of both A and C, but not dominated by properties characteristic of either A or C.

B & A   Any horizon qualifying as B in > 50% of its volume including parts that qualify as A2.

B   Horizons in which the dominant feature or features is one or more of the following: (i) an illuvial concentration of silicate clay, iron, aluminum, or humus, alone or in combination; (ii) a residual concentration of sesquioxides or silicate clays, alone or mixed, that has formed by means other than solution and removal of carbonates or more soluble salts; (iii) coatings of sesquioxides adequate to give conspicuously darker, stronger, or redder colors than overlying and underlying horizons in the same sequum but without apparent illuviation of iron and not genetically related to B horizons that meet requirements of (i) or (ii) in the same sequum; or (iv) an alteration of material from its original condition in sequums lacking conditions defined in (i), (ii), and (iii) that obliterates original rock structure, that forms silicate clays, liberates oxides, or both, and that forms granular, blocky, or prismatic structure if textures are such that volume changes accompany changes in moisture.

B1   A transitional horizon between B and A1 or between B and A2 in which the horizon is dominated by properties of an underlying B2 but has some subordinate properties of an overlying A1 or A2.

B2   That part of the B horizon where the properties on which the B is based are without clearly expressed subordinate characteristics indicating that the horizon is transitional to an adjacent overlying A or an adjacent underlying C or R.

B3   A transitional horizon between B and C or R in which the properties diagnostic of an overlying B2 are clearly expressed but are associated with clearly expressed properties characteristic of C or R.

C   A mineral horizon or layer, excluding bedrock, that is either like or unlike the material from which the solum is presumed to have formed, relatively little affected by pedogenic processes, and lacking properties diagnostic of A or B but including materials modified by: (i) weathering outside the zone of major biological activity; (ii) reversible cementation, development of brittleness, development of high bulk density, and other properties characteristic of fragipans; (iii) gleying; (iv) accumulation of calcium or magnesium carbonate or more soluble salts; (v) cementation by accumulation such as calcium or magnesium carbonate or more soluble salts; or (vi) cementation by alkali-soluble siliceous material or by iron and silica.

R   Underlying consolidated bedrock, such as granite, sandstone, or limestone. If presumed to be like the parent rock from which the adjacent overlying layer or horizon was formed, the symbol R is used alone. If presumed to be unlike the overlying material, the R is preceded by a Roman numeral denoting lithologic discontinuity.

soil map—A map showing the distribution of soil types or other soil mapping units in relation to the prominent physical and cultural features of the earth's surface. The following kinds of soil maps are recognized in the United States:

soil map, detailed—A soil map on which the boundaries are shown between all soil types that are significant to potential use as field-management systems. The scale of the map will depend upon the purpose to be served, the intensity of land use, the pattern of soils, and the scale of the other cartographic materials available. Traverses are usually made of 1/4-mile, or more frequent, intervals. Commonly a scale of 4 inches = 1 mile (1:15,840) is now used for field mapping in the U.S.

soil map, detailed reconnaissance—A reconnaissance map on which some areas or features are shown in greater detail than usual, or than others.

soil map, generalized—A small-scale soil map which shows the general distribution of soils within a large area and thus in less detail than on a detailed soil map. Generalized soil maps may vary from soil association maps of a county, on a scale of 1 inch = 1 mile (1:63,360), to maps of larger regions showing associations dominated by one or more great soil groups.

soil map, reconnaissance—A map showing the distribution of soils over a large area as determined by traversing the area at intervals varying from about 1/2 mile to several miles. The units shown are soil associations. Such a map is usually made only for exploratory purposes to outline areas of soil suitable for more intensive development. The scale is usually much smaller than for detailed soil maps.

soil map, schematic—A soil map compiled from scant knowledge of the soils of new and undeveloped regions by the application of available information about the soil-formation factors of the area. Usually on a small scale (1:1,000,000 or smaller). See **soil-formation factors.**

soil mechanics and engineering—A subspecialization of soil science concerned with the effect of forces on the soil and the application of engineering principles to problems involving the soil.

soil monolith—A vertical section of a soil profile removed from the soil and mounted for display or study.

soil morphology—(i) The physical constitution, particularly the structural properties, of a soil profile as exhibited by the kinds, thickness, and arrangement of the horizons in the profile, and by the texture, structure, consistency, and porosity of each horizon. (ii) The structural characteristics of the soil or any of its parts.

soil organic matter—The organic fraction of the soil; includes plant and animal residues at various stages of decomposition, cells and tissues of soil organisms, and substances synthesized by the soil population. Usually determined on soils which have been sieved through a 2.0-mm sieve.

soil population—All the organisms living in the soil, including plants and animals.

soil province—(Obsolete) Areas similar in mode of origin of the soil parent materials or in geological or geographic features.

soil salinity—The amount of soluble salts in a soil, expressed in terms of percentage, parts per million, or other convenient ratios.

soil science—That science dealing with soils as a natural resource on the surface of the earth including soil formation, classification and mapping, and the physical, chemical, biological, and fertility properties of soils per se; and these properties in relation to their management for crop production.

soil separates—Mineral particles, < 2.0 mm in equivalent diameter, ranging between specified size limits. The names and size limits of separates recognized in the United States are: *very coarse sand,* 2.0 to 1.0 mm; *coarse sand,* 1.0 to 0.5 mm; *medium sand,* 0.5 to 0.25 mm; *fine sand,* 0.25 to 0.10 mm; *very fine sand,* 0.10 to 0.05 mm; *silt,* 0.05 to 0.002 mm; and *clay,*[2] < 0.002 mm.

The separates recognized by the International Society of Soil Science are: I) *coarse sand,* 2.0 to 0.2 mm; II) *fine sand,* 0.2 to 0.02 mm; III) *silt,* 0.02 to 0.002 mm; and IV) *clay,* < 0.002 mm.

soil series—The basic unit of soil classification being a subdivision of a family and consisting of soils which are essentially alike in all major profile characteristics except the texture of the A horizon.

soil solution—The aqueous liquid phase of the soil and its solutes consisting of ions dissociated from the surfaces of the soil particles and of other soluble materials.

soil structure—The combination or arrangement of primary soil particles into secondary particles, units, or peds. These secondary units may be, but usually are not, arranged in the profile in such a manner as to give a distinctive characteristic pattern. The secondary units are characterized and classified on the basis of size, shape, and degree of distinctness into classes, types, and grades, respectively. Platy, prismatic, columnar, blocky or nut, subangular blocky (nut), granular, crumb.

---

[1]Prior to 1947 this separate was called "fine gravel;" now the gravel includes particles between 2.0 mm and about 12.5 mm in diameter.

[2]Prior to 1937, "clay,, included particles < 0.005 mm in diameter, and "silt," those particles from 0.05 to 0.005 mm.

soil survey—The systematic examination, description, classification, and mapping of soils in an area. Soil surveys are classified according to the kind and intensity of field examination.

soil texture—The relative proportions of the various soil separates in a soil. The textural classes may be modified by the addition of suitable adjectives when coarse fragments are present in substantial amounts; for example, "stony silt loam," or "silt loam, stony phase." (For other modifications see **coarse fragments.**) The sand, loamy sand, and sandy loam are further subdivided on the basis of the proportions of the various sand separates present.

soil type—(i) The lowest unit in the natural system of soil classification; a subdivision of a soil series and consisting of or describing soils that are alike in all characteristics including the texture of the A horizon. (ii) In Europe, roughly equivalent to a great soil group.

soil variant—A soil whose properties are believed to be sufficiently different from other known soils to justify a new series name but comprising such a limited geographic area that a creation of a new series is not justified.

solodized soil—A soil that has been subjected to the processes responsible for the development of a Soloth and having at least some of the characteristics of a Soloth.

Solonchak—A great soil group of the intrazonal order and halomorphic suborder, consisting of soils with gray, thin, salty crust on the surface, and with fine granular mulch immediately below being underlain with grayish, riable, salty soil; formed under subhumid to arid, hot or cool climate, under conditions of poor drainage, and under a sparse growth of halophytic grasses, shrubs, and some trees.

Solonetz—A great soil group of the intrazonal order and halomorphic suborder, consisting of soils with a very thin, friable, surface soil underlain by a dark, hard columnar layer usually highly alkaline; formed under subhumid to arid, hot to cool climates, under better drainage than Solonchaks, and under a native vegetation of halophytic plants.

solum (plural: sola)—The upper and most weathered part of the soil profile; the A and B horizons.

Subarctic Brown Forest soils—Soils similar to Brown Forest soils except having more shallow sola and average temperatures of $< 5C$ at 18 inches or more below the surface.

surface soil—The uppermost part of the soil, ordinarily moved in tillage, or its equivalent in uncultivated soils and ranging in depth from 3 or 4 inches to 8 or 10. Frequently designated as the "plow layer," the "Ap layer," or the "Ap horizon."

T

tight soil—A compact, relatively impervious and tenacious soil (or subsoil) which may or may not be plastic.

tilth—The physical condition of soil as related to its ease of tillage, fitness as a seedbed, and its impedance to seedling emergence and root penetration.

toposequence—A sequence of related soils that differ, one from the other, primarily because of *topography* as a soil-formation factor. See **clinosequence.**

topsoil—(i) The layer of soil moved in cultivation. See **surface soil.** (ii) The A horizon. (iii) The A1 horizon. (iv) Presumably fertile soil material used to topdress roadbanks, gardens, and lawns.

transitional soil—A soil with properties intermediate between those of two different soils and genetically related to them.

transported soil—(Obsolete) Any soil which was formed on unconsolidated sedimentary rocks.

truncated—Having lost all or part of the upper soil horizon or horizons.

Tundra soils—(i) Soils characteristic of tundra regions. (ii) A zonal great soil group consisting of soils with dark-brown peaty layers over grayish horizons mottled with rust and having continually frozen substrata; formed under frigid, humid climates, with poor drainage, and native vegetation of lichens, moss, flowering plants, and shrubs.

U

undifferentiated soil groups—Soil mapping units in which two or more similar taxonomic soil units occur, but not in a regular geographic association. For example, the steep phases of two or more similar soils might be shown as a unit on a map because topography dominates the properties. See **soil association** and **soil complex.**

urban land—Areas so altered or obstructed by urban works or structures that identification of soils is not feasible.

V

value, color—The relative lightness or intensity of color and approximately a function of the square root of the total amount of light. One of the three variables of color. See **Munsell color system, hue,** and **chroma.**

varve—A distinct band representing the annual deposit in sedimentary materials regardless of origin and usually consisting of two layers, one a thick, light-colored layer of silt and fine sand and the other a thin, dark-colored layer of clay.

volcanic-ash land—Areas of volcanic ash so recently deposited that the ash is essentially unmodified and shows little or no evidence of soil development. The areas have almost no vegetation on them.

W

water table—The upper surface of groundwater or that level below which the soil is saturated with water; locus of points in soil water at which the hydraulic pressure is equal to atmospheric pressure.

water table, perched—The water table of a saturated layer of soil which is separated from an underlying saturated layer by an unsaturated layer.

weathering—All physical and chemical changes produced in rocks, at or near the earth's surface, by atmospheric agents.

Z

zonal soil—(i) A soil characteristic of a large area or zone. (ii) One of the three primary subdivisions (orders) in soil classification as used in the United States. See **classification, soil.**

Index